POLITICAL EXERCISE

POLITICAL EXERCISE

ACTIVE LIVING, PUBLIC POLICY, AND THE BUILT ENVIRONMENT

LAWRENCE D. BROWN

Columbia University Press *New York*

Columbia University Press
Publishers Since 1893
New York Chichester, West Sussex
cup.columbia.edu

Library of Congress Cataloging-in-Publication Data
Names: Brown, Lawrence D., author.
Title: Political exercise : active living, public policy, and the built environment / Lawrence D. Brown.
Description: New York : Columbia University Press, 2022. | Includes bibliographical references and index.
Identifiers: LCCN 2021029920 (print) | LCCN 2021029921 (ebook) | ISBN 9780231173506 (hardback) | ISBN 9780231173513 (trade paperback) | ISBN 9780231553445 (ebook)
Subjects: LCSH: Public health—Political aspects—Case studies. | Public health—Social aspects. | Health planning. | City planning—Health aspects.
Classification: LCC RA425 .B7724 2022 (print) | LCC RA425 (ebook) | DDC 362.1—dc23
LC record available at https://lccn.loc.gov/2021029920
LC ebook record available at https://lccn.loc.gov/2021029921

COVER DESIGN: Milenda Nan Ok Lee
COVER IMAGE: Patti McConville © Alamy

To Adela, Madeleine, and Tristan

CONTENTS

ACKNOWLEDGMENTS

As a student of health care policy I understood long ago that the promotion of physical exercise stands very high among strategies for improving individual and population health, but I had not much engaged with it in my research. This gap was filled thanks to the Robert Wood Johnson Foundation, which went beyond the usual conceptual and institutional suspects in health policy by creating, in 2001, an innovative program called Active Living by Design (ALbD) that funded twenty-five U.S. communities to pursue changes in their built environments that encouraged walking, biking, and other forms of "active living," and in the evaluation of which I participated. M. Katherine (Kate) Kraft, who took the lead in designing and implementing the program, has been a superb mentor and endless source of insights on active living, both during her time at RWJF and, more recently, as a consultant and executive director of America Walks. I am also grateful to Sarah Strunk, who directed the National Program Office of ALbD at the University of North Carolina during the years of my research and helped me in countless ways.

Between 2007 and 2011 I made field trips to four of the twenty-five funded sites (Wilkes-Barre, Louisville, Albuquerque, and Sacramento) and then gave the foundation a report of ninety pages. Uncertain whether to shrink that document into article size or expand it into a book, I chose the latter course and widened my focus to examine the progress of active living projects and policies in five communities (the original four plus New York City), tracking and analyzing plans and interventions that extended beyond both the life of the ALbD grants and the scope of the projects they supported. My work depended crucially on roughly 100 in-person interviews (most one-on-one, a few in small groups) with local sources in a range of sectors (health policy, transportation, public works, city planning, parks, schools, and more) who generously shared their time and insights with me and supplied useful documents. I am particularly grateful to Michele Schasberger, Carol Hussa, Rolf Eisinger, Cassandra Culin, Cheri Bryant Hamilton, Isaac Benton, Tara Cok, Anne Geraghty, Teri Duarte, Ted Link-Oberstar, Lynn Silver, and Lorna Thorpe, who kindly read and commented on individual chapters. They, of course, bear no responsibility for such errors of fact or interpretation as the book may contain.

Thanks as well to Michael Sparer and my other colleagues in the Department of Health Policy and Management in the Mailman School of Public Health at Columbia University, who endured my ruminations on active living in two brown-bag seminars and in many informal conversations. Anne-Laure Beaussier of Sciences Po organized a seminar at which I presented my research in 2018 at the Centre for the Sociology of Organisations in Paris. Three other French colleagues—William Genieys, Laetitia Atlani-Duault, and Michel Naiditch—also offered valuable comments on my clumsy stabs at sense-making. Phoebe Downer rendered excellent research assistance.

Columbia University Press has been a pleasure to work with at each step of the publication process. Patrick Fitzgerald got the manuscript up and running. Lowell Frye offered excellent counsel as it took shape and upon its completion enlisted two anonymous reviewers whose suggestions markedly improved its quality. Patricia Bower expertly edited it, and Kara Pekar created the index.

My warmest thanks go to my wife, Adela; my daughter, Madeleine; and my son, Tristan, who bemusedly tolerated my musings on sidewalks and bike lanes and gently exhorted me to get a move on with the writing.

INTRODUCTION

We must not begin by talking of pure ideas,—vagabond thoughts that tramp the public roads without any human habitation,—but must begin with men and their conversation.

—Charles Sanders Peirce

That active living—physical exercise, preferably integrated into everyday routines—brings impressive health and social benefits is an ancient and durable notion whose time has come again with special force in the twenty-first century. Celebrated since Before the Common Era as a source of martial virtue, fun and games, psychological equipoise, and bodily vigor, the many forms of physical exertion that constitute "active living" have come front and center today as policymakers seek not only to slow the rising costs of chronic illness but also, and more broadly, to create sustainable metropolitan configurations.

Both as a strategy for improving the health of the public and as an element in movements such as New Urbanism, smart growth, environmental protection, and sustainable development, active living comes as close to commanding consensus as one could expect among these vexing ventures in social (re)engineering. Yet, as often happens in public affairs, how to get from here (a widely accepted call for public intervention) to there (a set of projects, programs, policies, and systemic changes that work well in a range of settings and for a broad swath of the population) remains conceptually and strategically elusive. The pages that follow argue that the key problem afflicting policy efforts to promote active living is the failure in public health and social policy circles to confront an uncomfortable truth—namely, that effective promotion of active living is a pervasively political exercise, the workings of which remain today clouded by mystery and, indeed, paradox.[1] This book aims to illuminate this strategic terrain and thereby to help policymakers and other would-be reformers to reap the benefits of active living in both the theory and, more important, the practice of contemporary public problem solving. Evidence may speak for itself, but it does not act by itself.

Those who hold that health policy should be a faithful implementation of clinical or epidemiological versions of "doctors' orders" may recoil to read that politics is essential to achieving the positive health outcomes at which authoritative professional instructions aim. But even a cursory inspection of U.S. health policies—for example, biomedical research in and under the National Institutes of Health, hospital construction by means of the Hill-Burton Act, workforce training programs, Medicare, Medicaid, the Children's Health Insurance Program, the Affordable Care Act, the promotion (and regulation) of managed care, and many more—reveals the omnipresent politics of agenda-setting, formulation, implementation, and refinement that launched these initiatives and sustained (or modified or terminated) them over the years. Policies can be more or less infused with evidence, but there can be no such thing as an apolitical health policy or program.

The centrality of politics is also plain in policy venues that work to improve the health of the public by means of interventions that reach beyond the above-mentioned incursions into medical care and coverage. A case in point is the "doctor's order" of the Office of the U.S. Surgeon General that health-seeking citizens should walk briskly for thirty minutes or so several days each week. Those walks presuppose intermediate steps that are shaped by myriad political processes. Accessible and navigable sidewalks come into place (or not) because municipal departments of public works use funds, appropriated by elected local officials, to put them here and there or because public zoning codes require private developers to install them.

The speed and diligence with which sidewalks are maintained and repaired may reflect the political clout of neighborhoods. The degree to which residents find them safe and inviting to walk on depends partly on how well lit and well policed they are—and perhaps too on whether they lead to appealing destinations, which depends in turn on the politics of city planning and design. Some would-be walkers may figure it out for themselves, but others may be mobilized by encouragement from city hall and by other public voices (mayors' miles and marches and such). Parks, trails, and recreation facilities do not spring spontaneously from private sector soil; rather they require public authority and capital for construction, maintenance, and staff.

Bike lanes follow similar strategic routes. Each stage along these routes—getting active living initiatives on the agenda of actors with influence; enactment of public laws, regulations, codes, and guidelines; allocation of (and tradeoffs among) public funds; implementation of plans—takes shape "internally" in the hands of a range of public officials, elected and appointed, and responds at every step to demands of attentive, activated groups (pedestrians, cyclists, contractors, home builders, neighborhood associations, and "urbanists" both new and old, for example) that are formally "external" to the political process.

These internal and external actors and processes engage, too, with private market forces (and, at times, with policy analysts) to generate the built environment, the materially and socially constructed composition and character of the common space in which walkers march to the surgeon general's orders. Public health and other scientific authorities eager to promote active living (or other "lifestyle" changes conducive to better health of the public) from above, as it were, might care to ponder Harry Truman's famous prediction of the dispiriting fate of orders enunciated by "Poor Ike," his presidential successor, former general Dwight D. Eisenhower: "He'll sit here and he'll say 'Do this! Do that!' And nothing will happen."[2] Making active living happen is the province of politics—and the theme of this book.

ORIGINS AND ORIENTATIONS

Active living is a compendium of themes that are deeply rooted, widespread, persistent, overlapping, and interactive. Five such themes stand out.

Martial Virtue. In ancient Greece and the Roman Empire, gymnastics and other physical training were highly valued as means to victory in military combat, and fitness for battle has been a recurring spur to the official pursuit of active living by national states through the years.[3] The victory of Germany over France in 1870, for example, showed that "German soldiers were physically stronger and healthier than their French counterparts" and lent intensity to the Germans' fascination with gymnastics—and with broader health care and coverage.[4] (The French, meanwhile, preferred bicycling.[5]) In the United States, statistics showing the sizable number of draftees who were deemed unfit for service in World War I and then again in World War II prompted proposals to expand physical education classes in the nation's schools.[6] Apparently much work remains to be done: a report by the Council for a Strong America recently warned that a disturbing number of potential military recruits are "Unhealthy and Unprepared" as a consequence of obesity and other lifestyle-related conditions.[7]

Fun and Games. Physical activity is integral to the "play element," an "absolutely primary category of life" that throughout history has been expressed in the "great competitions . . . [that were] indispensable as health and happiness-bringing activities."[8] And the import of these pursuits far exceeds what the vernacular calls "mere" fun and games. The exercise of "keeping together in time"—dance, calisthenics, gymnastics, marching, and kindred activities that entail "prolonged and insistent rhythmic stimuli"—contributes powerfully to social integration, to the "rise of civilized complexity" and to solidarity within communities and nations.[9] These appeals stretch, too, into rarified strata of society. Thus, around the time of World War I Marcel Proust wrote that a "feverish intensity" (*une vivacite febrile*) accompanying "the favor enjoyed by physical exercises" was supplanting the nonchalance of men heretofore accustomed to idleness.[10] In 1933 the Athens Charter adopted by the International Congress of Modern Architecture enshrined "recreation and leisure" (with a focus on sport, parks, and stadia) as one of the four basic functions to which any city should attend.[11]

Folk Wisdom. The virtues of active living have long been captured in a conventional wisdom that honors the behavioral concomitants of sound bodies and minds. Galen of Pergamum praised the power of moderate exercise—movement that causes its practitioner to "breathe more or less faster"—because it "renders the pulse vigorous, large, quick, and frequent," and indeed "has the capacity to provide health of the body, harmony of the parts, and virtue in the soul."[12] Maimonides prescribed "physical gymnastics to the point that the soul . . . rejoices." Søren Kierkegaard remarked how he daily walked himself "into a state of well-being and . . . away from every illness," and deplored the sedentary life: "the more one sits still, the closer one comes to feeling ill." The playground movement launched by Hull House in the 1880s in Chicago contended that "engaging children in active recreation and organized play was . . . a way to deflect them from the many illicit temptations of the city."[13] Late-nineteenth-century Germans found in regular exercise ways to improve the body, avert tuberculosis, and control "lascivious thinking."[14] Their Weimar successors cherished "healthful walks," while Nazi competitors sought legislation providing for "compulsory gymnastics and sports" and support for "all organizations engaged in the physician training of youth."[15] In the Gibbsville of John O'Hara, the precept "walk a mile a day" reigned in the code of healthy living alongside "[drink] a cup of hot water in the morning, Don't Worry, take a nap every afternoon . . . The Golden Rule [and] visit your dentist twice a year."[16]

Doctors' Orders. Nowadays health benefits of active living such as joyful souls, chaste thoughts, and clear lungs take a back seat to more precise clinical considerations—for instance, how a sustained elevated heart rate may improve cardio-pulmonary functions, promote the assimilation of glucose by the muscles, and reduce blood pressure. Public leaders, among whom Theodore Roosevelt is an iconic case in point, have touted the physical blessings of exercise for more than a century. Dwight Eisenhower created in 1956 a President's Council on Youth Fitness, a body that John F. Kennedy broadened into a President's Council on Physical Fitness in 1963. In 1968 Lyndon Johnson inaugurated a Presidential Fitness Award and enlarged the council's remit to physical fitness and sports. (In 2010 nutrition was added to its title.)

In the late twentieth century authoritative medical sources, including the U.S. surgeon general, found evermore numerous occasions to extoll the benefits of active living and to offer increasingly detailed directives for individuals seeking to attain them. The drumbeat quickened in and after 1995, the year in which the federal Centers for Disease Control (CDC) joined with the American Academy of Sports Medicine to offer quantified recommendations for exercise.[17] In short order the U.S. Surgeon General's office published a report, *Physical Activity and Health*, bristling with numerical targets of its own, and in 1996 thirty health-centered organizations created a National Coalition for Promoting Physical Activity.[18] In 1999 the CDC created its Division of Nutrition, Physical Activity, and Obesity.

In the first two decades of the twenty-first century, attention to the physiological virtues of physical exercise intensified. In 2008 the surgeon general reiterated and refined guidelines that urged adults to engage in at least 150 minutes per week of moderate-intensity aerobic activity or 75 minutes of vigorous-intensity exertion in order to enjoy "substantial health benefits." Children and adolescents, the guidelines counseled, should aim for one hour of physical activity daily, and the public was urged to supplement aerobic activities with muscle-strengthening exercises on two or more days each week.[19] In 2009 the *New York Times* launched a new column on physical fitness in the Science section of its Tuesday edition. In 2010 the Affordable Care Act established a Prevention and Public Health Fund and required insurers to cover "physical activity counseling," among other services recommended by the U.S. Preventive Services Task Force. Meanwhile, First Lady Michelle Obama made her own hortatory contribution to the pursuit of active living, a "Let's Move" program. In 2011 the National Prevention Council (composed of seventeen federal agencies and chaired by the surgeon general) issued its "National Prevention Strategy," two of the forty-one pages of which are devoted to active living as a "priority." In 2013 the august Institute of Medicine joined the chorus with its report *Educating the Student Body: Taking Physical Activity and Physical Education to School.*[20]

Addressing Chronic Conditions. The fifth and most capacious variation on the score that sings the benefits of active living views it as crucial to forestalling and managing chronic health conditions. Management of chronic illness is arguably the preeminent challenge facing U.S. health care policy today. The problem, which results from the happy twentieth-century conjunction of the conquest of infectious diseases (for which public health may properly claim much credit) and the manifold fruits of medical innovation, is that unprecedentedly large numbers of people (especially ones blessed by residence in affluent zip codes) can now expect to live at least into their nineties, although perhaps afflicted with morbidities, often multiple, that would have taken years, perhaps decades, off their lives before. If these additive demographic circumstances (more people living with and seeking care for more medical conditions over longer life spans) are addressed mainly by additive medical responses (more visits to physicians, more highly technological interventions, more admissions to hospitals, and so on) by an acute care system that, in the United States, famously equates more with better, then proper access to high quality care may be hard to maintain (not least for chronic care patients themselves), and of course health care costs—already high—will soar.

How best to organize the system to manage current and future demand for chronic care is a question that pervades the human and health policy life cycles from, as it were, cradle to grave. A growing end-of-life literature ponders how to constrain the semireflexive application of advanced medical treatments, how to ease the way to natural death for patients who request it, and how to define defensible "pathways" for the dying. Meanwhile, in the course of "normal" care, policymakers work to develop new integrative settings, such as medical homes and accountable care organizations, that mobilize teams of professionals (general practitioners, specialists, nurses, therapists, nutritionists, and social workers, for example) to manage efficiently the care of chronic care patients (mainly meaning with less resort to emergency rooms and inpatient beds). And the hope—increasingly backed by research evidence—that the fiscal and physical pain accompanying decades of life with debilitating conditions can be mitigated by preventive and coping measures earlier in the life cycle shines a fresh and favorable light on remedies that the public health sector might supply not only for crises du jour but also for *la longue durée*. The onset and worsening of ailments such as hypertension, diabetes, asthma, congestive heart failure, stroke, and cancer may be delayed, perhaps prevented, by lifestyle choices—refrain from smoking, avoid overuse of drugs and alcohol, eat healthy foods, *and* do regular physical exercise (that is, active living), which stands high on virtually all lists of sound policies.

The poster child for supposedly avoidable chronic conditions is of course obesity, and the central role ascribed to active living in combating the so-called epidemic of obesity has much amplified its appeal as a preventive strategy. This means–end coupling is problematic, however. First, designating obesity as a disease, let alone an epidemic, is questionable. People do not die *of* obesity but rather of conditions with which it may, but does not invariably, correlate. Second, weight and body mass index are crude indicators of physical health. And third, whatever benefits the war on weight may realize are counterbalanced by serious costs—for instance, the stigmatization of the overweight, the inability of most people who lose weight to keep it off, and the resort to dangerous drugs and risky crash diets.[21] The case for active living has nothing to do with obesity per se. Research suggests not merely that "exercise itself is a major determinant of . . . health," independent of loss of weight, but also that inactivity, not obesity, "may be the real chronic health problem that America faces."[22]

ACTIVE LIVING AS POLICY PARADOX

"Paradoxes," writes Deborah Stone, "are nothing but trouble. They violate the most elementary principle of logic. Something cannot be two different things at once."[23] Three paradoxes haunt the transition of active living policy from theory into practice, and they are troubling indeed.

The first paradox appears within the realm of public and professional opinion. As a precept—indeed, as a staple in a modern design for a healthy life—active living comes as close to commanding consensus as anything floating in the policy ether. Virtually no one dissents from the proposition that more regular physical exercise among the population would be a good thing, or objects to the general proposition that public and private leaders should encourage it by means of education, exhortation, incentives, and changes in the built environment. Both the base of evidence that documents the benefits of active living and the ranks of enthusiasts who want it to hold a higher priority in public policy have expanded steadily.[24]

Paradoxically, however, progress by Americans toward attainment of the goals for activity, such as those promulgated by the surgeon general, continues to stagnate. In the mid-1990s, more than 60 percent of American adults and nearly half of youth of aged twelve to twenty-one did not pursue regular physical activity; 25 percent of adults were "not active at all."[25] Almost twenty years later, the needle had barely moved. For example, among children aged six to eleven, 57.8 percent of males and 67 percent of females had no "daily vigorous physical exercise" in 2007. In 2011 a mere 21 percent of Americans aged eighteen and above "met both aerobic activity and muscle-strengthening guidelines," and nearly half (47.6 percent) met neither guideline.[26] One survey found that Americans average only about half the daily steps recommended for those who hope to gain health benefits from walking, and that on this count Americans fell short of the Swiss, Australians, and Japanese.[27] Apparently a disconnect persists between the theoretical appeal of active living and the practical production of both enlarged demand for it and a larger supply of settings that encourage it.

A second paradoxical feature of active living policy derives from organizational life. On the one hand, encouragement to physical exercise shows up on the mission statements and strategic agendas not only of groups explicitly devoted to the promotion of walking and cycling but also on those of a long list of movements (smart growth, New Urbanism, environmental protection, and sustainable communities, for example), voluntary organizations (YMCAs, urban leagues, parent-teacher associations), and public agencies (federal, state, county, and local departments of health, transportation, public works, education, parks and recreation, and environmental protection)—a pattern that bespeaks a broad coalition and the political clout that goes with it. On the other hand, with the obvious exception of organized bikers and pedestrians, active living is not a top priority within the missions and practical agendas of any of these entities. To lift an image from Kenneth Burke's dissection of another sort of paradox, organizational support for active living has an extensive circumference but no solid center.[28]

The importance of the Active Living by Design (ALbD) program, launched by the Robert Wood Johnson Foundation in 2002, was precisely to create such a center. Between 2003 and 2009 the program gave five-year grants to twenty-five communities charged with developing programs and policies that could help to "reengineer activity back into people's lives."[29] But getting the center to hold over time has not been easy. Alliances that look solid on the surface may encounter, within and between constituent groups, changing priorities that reflect the capacities of these groups to attract and confer resources and influence. The funds and prestige of the ALbD grant, for instance, elevated the status of proponents of active living in the eyes of (among others) advocates for smart growth, New Urbanism, and sustainable communities, who agreed to put their shoulders to the wheel in pursuit of common goals and innovative projects. But when grants end and local projects become yesterday's news, allies tend to focus anew on their core missions, leaving active living aficionados wondering what to do for an encore. Lifting the status of active living from latent to manifest within and across the populous organizational universe the contributions of which are key to making it "happen" and "work" is no easy task. The evolution of ALbD within the Johnson Foundation ironically illustrates the paradox of organizational priority. No sooner had it begun launching projects in its funded twenty-five sites, the program lost some of its distinct programmatic identity when in 2003 a new president arrived at the foundation, refocused it on "preventing childhood obesity," and put it under the broader umbrella of its Childhood Obesity Team.[30]

Paradox number three besets the strategic options and policy levers available to political leaders who champion active living. Although the basic "technology" of active living is uncomplicated, how to apply it effectively by means of public policy remains in many ways mysterious. Communicating to the citizenry the benefits of brisk daily walks and more bike lanes on roadways does not entail the kind of intricate scientific adventures that launch rockets and cure cancer. Yet getting the active living message to register and reshaping the built environment for the benefit of walkers and cyclists repeatedly proves to be much harder than any logic of rational problem solving would seem to predict.[31]

For active living, as for many other important population-based interventions, the public health message is straightforward: policies that discourage smoking, substance abuse, road accidents, unsafe sex, unhealthy diets, and sedentary lifestyles prevent the incidence of some diseases, limit the onset and severity of others, mightily improve the quality of life of much of the citizenry, and lower the growth of health care costs. And sometimes the medium (that is, the set of public tools to be deployed) is as translucent as the message. For instance, a combination of information, regulation, and (especially) taxes has lowered rates of smoking, and tougher enforcement of laws against drunk driving has reduced fatalities from auto accidents.

Clear policy goals, however, are not always accompanied by dependable and acceptable means of intervention. The tested strategic trio of information, regulation, and taxes does not transfer seamlessly to active living, and the progeny of behavioral economics, such as personal financial incentives—payments to individuals who lose weight, for example—seem to be markedly less effective in this sphere than at motivating one-off preventive practices such as immunizations.[32] Because workable strategies to promote active living cannot be lifted off a theoretical shelf, political champions must cobble them together from bits and pieces of support within and across communities, policy sectors, and intergovernmental levels.

PARADOX LOST? ACTIVE LIVING AND SUSTAINABLE SOCIETIES

Arguably there is nothing so very paradoxical about these failed connections between the abstract adherence to the precepts of active living and the practical capacity of U.S. society to honor them. Even a cursory inspection of the political economy of metropolitan development in the United States would seem to lay these paradoxes to rest. American society has for decades acquiesced—indeed, indulged—in patterns of regional growth that have engineered active living out of daily life and have created in the process built environments that are wasteful of natural resources, conducive to social isolation, and unhealthy to boot.

By the 1920s cars had become "an accepted essential of normal living," and metropolitan sprawl, a feature of urban life in the United States since the advent of "streetcar suburbs" in the late nineteenth century, accelerated markedly after World War II with the aid of public policies that responded to the revealed residential preferences of consumers.[33] Policymakers acted, then and since, on the premise that homebuyers (and some renters) were and remain eager to flee the spatial density, social heterogeneity, and heterodox land-use patterns of traditional center cities and to seek instead single-family homes that occupy a plot of land with space for yard, driveway, and garage; are surrounded by other homes that are broadly similar in design and market value; have neighbors of similar class, racial, and perhaps ethnic characteristics; and are cordoned off from discordant commercial and industrial enterprises. Federal mortgage assistance policies and the GI Bill for returning members of the armed forces complemented federal support for the construction of highways, and even when mass transit systems were part of the picture, the suburban layout in U.S. metropolitan areas that channeled major population growth in the second half of the twentieth century depended heavily on cars (and trucks) and therefore on roadways that were expected to carry these vehicles quickly, conveniently, and safely.

These deep-seated spatial arrangements have produced built environments that are on many counts antagonistic to active living. In a society that prized "speed, and space, and privacy," by the 1960s, "to consider walking anywhere" seemed to be "suspect and even un-American."[34] Indeed the situation has "given rise to a joke: in America, a pedestrian is someone who has just parked their car."[35] Would-be walkers find no sidewalks, or ones in disrepair (which can cause falls, especially by the aged); routes interrupted by broad avenues and tricky crossings amid heavy traffic; traffic that is too close for comfort; and, not least important, nowhere interesting to go (that is, no inviting downtown destinations with shops, cafes, theaters, and other attractions in close proximity to each other). In many residential communities the simplest errand—buying milk, eating out—means taking the car because zoning rules discourage "mixed use."

Children seldom walk or bike to school because distances are too great, traffic along the way is too heavy, and parents are fearful that their kids may be abducted or assaulted by strangers. In 1969 nearly half of the nation's school children walked or biked to school.[36] In 2014 a study of CDC data found that in 61.5 percent of schools, only 10 percent of students or fewer did so on an average school day, and in only 22.7 percent of schools did 26 percent or more students walk or bike.[37] Even the solitude and splendor of suburban trails, paths, and parks present complications—for example, poorly maintained surfaces that make walking hazardous and a lack of parking for people who live too far away to walk to their preferred walking site.

Bicyclists, too, are encumbered. Riding on sidewalks makes them a menace to pedestrians; riding on roadways pits them against cars and trucks. Bike lanes, when available, are often narrow. Pleasant, sustained biking requires continuity, hence "connectivity" of routes so as to avert frequent stopping, dismounting, and crossing busy thoroughfares. Scarcity of bike racks and other secure places in which to store bikes deters riders from using them to travel to work (or to the train) and complicates their maintenance when they cannot be easily brought into and out of apartments.

The engineering of local space often ensures that the most appealing and feasible options for daily life invite inactivity: stay home or take the car. For many Americans, then, the daily routine involves a commute to work (often on congested highways), a day spent at a computer (or desk or counter), a car ride back home (with perhaps a stop at the mall for supplies or at a fast-food outlet for snacks), and an evening in front of a television or computer. Not a pretty sight, but straight down to the closing years of the twentieth century one would be hard pressed to deny that the people had consistently voted with their feet (firmly affixed to the gas pedal) in favor of metropolitan sprawl. The quest for more active-living-friendly built environments is therefore a struggle less with the physical laws of engineering than with cultural patterns that these physical structures express and with the myriad economic interests served by sprawl. For decades public policy has sustained, shaped, honored, and reinforced these interfusing patterns and interests. Arguably, then, the paradox of active living dissolves upon a clear-eyed confrontation with power—or, better, with the powers that drive the political economy of metropolitan development in the United States.

This developmental juggernaut is no longer the whole story, however. Critics have long assailed the forces sketched here—for their destruction of "urban villages," for instance, for their devaluation of "lost" inner city communities, and for their principled and planned antagonism to the vibrant, mixed-use neighborhoods that urban enthusiasts think essential to the greatness of American cities.[38] As social costs of urban renewal gained attention, the "trade-off between urban improvement and authentic locality" grew sharper.[39]

In the 1960s diffuse nostalgia and regret increasingly crystalized into organized political opposition to the "mega-projects" that ravaged cherished urban terrain.[40] Protests against destructive highway projects in New Orleans and other cities made plain the "enormous power of local and loosely-knit organizations."[41] The Highway Action Coalition, intent on "busting" the federal Highway Trust Fund, began challenging the heretofore impregnable "road gang" lobby and scored a victory in 1975 when Congress agreed to open the fund to support the construction of mass transportation projects, thus signaling that highway planners must henceforth proceed with a "very different set of assumptions and guidelines."[42] In 1991 the Intermodal Surface

Transportation Efficiency Act further expanded the range of options eligible for trust-funding.

These "pro-urban" arguments and appeals to "alternative transportation" once seemed to swim feebly against the tides of the U.S. metropolitan political economy, but in the early twenty-first century they have ceased to look antiquarian and elegiac. Reactions to the built environment that public policy fostered now flow with increasing force from multiple sources, the convergence of which couples powerful critical streams with imposing policy prescriptions for change.

The environmental protection movement that gained strength in the late 1960s deplored the air pollution that accompanies heavy reliance on automobiles in areas where neither workplaces, shops, nor schools are conveniently accessible for walkers and cyclists, and rising anxiety over climate change and global warming in the twenty-first century has reinforced the commitment to cleaner air. The oil price shocks of the mid-1970s and intermittent spikes in gas prices thereafter have underscored the financial costs of lifestyles predicated on long and frequent journeys by car. Proponents of smart growth and New Urbanism expatiate on the virtues of dense, mixed-used urban developments—the savings in time and money, the overcoming of social isolation and segregation, the rich cultural stimuli of authentic urban life. Cheaper airfares have enabled millions of tourists to see and enjoy such urban paragons (mostly outside the United States) for themselves.

The predilections of growing numbers of "singletons" who seek milieux in which to mix and mingle easily before (or instead of) settling down to family life have augmented demand for housing in compact, complex urban configurations.[43] And of course rising attention to the adverse health effects of suburban arrangements that consign residents to long sedentary hours in cars and leave little time and energy for physical exercise has triggered new interest in walking and biking and has emboldened local, state, and national associations of pedestrians and cyclists —including political action committees[44]—to lobby for policies that make metropolitan built environments more hospitable and attractive to their constituents. In short, as the twenty-first century dawned, active living was transcending its traditional rationales to become a highly active ingredient in a larger design for living that substantial segments of the population deemed healthier and more sustainable not only physiologically but also from social, economic, environmental, aesthetic, cultural, and psychological vantage points.

This new multiplex imagery of the built environment and of its connections to active living has thrown down a gauntlet before policymakers and stakeholders in the private and voluntary sectors and at all levels of the U.S. public sector. Simple cross-national comparisons show how much work the nation has in store. In a recent study Ralph Buehler and John Pucher report that the United States surpasses its Western peers in ownership of cars and light trucks (37 percent higher than in Germany, for instance), in use of cars (87 percent above the German rate), and in the percent of all trips that are made by cars.[45] (In the United States the breakdown among car trips, walking, and biking is 86 percent / 11 percent / 2 percent versus 66 percent / 24 percent / 10 percent in Germany.[46])

Germany has less than half the number of traffic fatalities per one hundred thousand residents as does the United States, and that disparity remains large even when one controls for differences in volume of use of cars. Safety of pedestrians is five times higher, and that of cyclists three times higher, in Germany than in the United States.[47] Transport costs consume an average of 17 percent of the disposable income of U.S. households and

14.6 percent of those in Germany (a difference of roughly $2,500 per year). The superior fuel efficiency of German cars and light trucks (52 percent above that in the United States) combined with lower mileage driven enables Germans to consume only one-third as much energy per capita per year as does the United States for transport, a sector that accounts for 30 percent of CO_2 emissions here and only 20 percent in Germany.[48] No less arresting than these numbers, moreover, is the ease with which they can be arrayed on one page in a single chart framing the social, economic, and environmental concomitants of "transport sustainability" in an analytical gestalt that captures much of the "design for living" emerging in the social and policy spheres.[49]

THE POLITICS OF PARADOX

The policies that shape metropolitan development and the character of the built environments that evolve therein are themselves encased in paradox. On the one hand, the physical and economic development of U.S. cities has been "driven overwhelmingly . . . by private for-profit investment" and by competition "to make themselves attractive to private investors."[50] On the other hand, "All action to do with planning and building is inevitably political," a product of "political agency."[51] Political agents willy-nilly choose a mission and role vis-à-vis the private drivers of urban development along a range from faithful reflection of private interests, to mediation between those interests and those of a larger public, to leadership that seeks to shape the definition and steer the direction of private interests. Those choices contribute to the construction of "physical fabrics" that differ markedly across space and time.[52]

The policy paradoxes of active living, nested within these encompassing tensions of political economy, embody a pervading political predicament of their own—one that finds generational change mediating between competing images of a desirable built environment and of the degree to which public policy should actively encourage physical exercise in the daily life of citizens. The traditional image, which honors sprawling suburban development, resistance to mixed land use, and heavy reliance on cars and the roadways on which they move, enjoys the powerful political support of developers, real estate agents, lenders, contractors, road builders, and elected officials at the federal, state, county, and local levels of government who share their views and appreciate their campaign contributions.

The challenger, which insists on the virtues of deliberative metropolitan planning, density, land use that encourages the mixing and mingling of diverse enterprises and activities, and spatial arrangements that discourage cars in favor of pedestrians, cyclists, and mass transit, gains influence among millennials, New Urbanists, advocates of smart growth, environmental protectors, climate change activists, advocates for greenness and sustainability, residents who elect to walk or bike for pleasure or health improvement, and political leaders (increasingly in suburbs as well as in cities) who endorse or respond to the growing chorus of progressive voices within the communities they represent. The paradoxes that captured the weakness of active living policy in the second half of the twentieth century are therefore by no means set in stone but rather stand subject both to reformulation by (to recall Arthur Bentley) the pulling and hauling of countervailing powers energized by interests and ideas and to accommodations that are renegotiated and reconfigured over time by the dynamics of generational change. In short, the promotion of active living is a perennial political exercise.

POLITICAL EXERCISE AND RESEARCH PRACTICE

Clifford Geertz recalled Michel Foucault's observation that "we usually know what we are doing, we sometimes know why we are doing it, but we almost never know what our doing does."[53] The pages that follow shed light on all three of these dimensions by examining how the political exercise that is the promotion of active living, by modifying the built environment in ways that encourage physical exercise, has fared in five U.S. cities early in the twenty-first century.

The five sites explored here were chosen because, taken together, they embody variation by size, region, socioeconomic mix, strategy to promote active living, and state context. Wilkes-Barre, Pennsylvania, a small "Rust Belt" city in the Northeast sought both to create a more walkable downtown and to expand and improve trails in its outskirts. Louisville, Kentucky, is a medium-large city at the heart of a metropolitan government in a southern state and is instructive in part for its early attempts to promote active living in lower-income housing projects. Albuquerque, New Mexico, the largest city in a Sunbelt state with a sizable number of Latinos and Native Americans, labored to promote "density" in a cultural landscape traditionally indulgent of "sprawl." On the West Coast, Sacramento is the capital city of California, at once the nation's largest state and the one that has pushed hardest for clean air, a quest that has made it uncommonly supportive (at least in principle) of "active transportation."

These four cities were selected from the twenty-five communities funded by the above-mentioned Active Living by Design program of the Robert Wood Johnson Foundation, to the evaluation of which this study contributed. The fifth site, New York

City, the nation's most populous and highly heterogenous, was added because it made impressive advances on several fronts, even in the face of many physical, political, and bureaucratic obstacles to active living.

Doubtless other samples of sites could have been equally instructive, but there is no "scientific" way to identify them in advance. The objective of the research was first and above all to discover "what goes on around here" when communities endeavor to change their built environments in ways that conduce to more active living—an aim that requires avoiding the preconceptions inherent in formal hypotheses that might (mis)guide the choice of sites. Rigid methodological maximizing yields here (as it should) to a rough and ready satisficing process of mixing and matching.

The methods employed in this research are unapologetically simple and eclectic. Review of published and online literature identified the main protagonists in the active living "field" and in the four funded sites. A (manageable) number of these actors were interviewed and asked in turn for suggestions of others who were central to, or well informed about, the design and progress of the local active living projects. Between 2008 and 2018, three rounds of interviews in each of the four cities were conducted on site with officials in and observers of the health care sector (departments of public health, hospitals, insurers), departments of transportation and of public works, city planners, schools, mayors' offices, city and county councils, metropolitan planning organizations, trail associations, and other institutional venues. When New York City was added as a fifth site, a comparable range of actors was interviewed. Most interviewees kindly supplied pertinent documents that detailed aspects of their efforts to make their built environment more hospitable to active living.

How then to proceed? The political exercise that is the promotion of active living plays out in communities across the country, but little is known about its structures, processes, and outcomes. Neither the social sciences nor the literatures of public policy and public health offer much by way of useful prefabricated theoretical templates to guide an exploration, and (no less than in case of site selection) imprisoning the raw material within hypotheses that foreshorten and constrain a researcher's gaze and imagination would be absurd—indeed, unscientific.

The sole helpful methodological hint is "to penetrate deeply enough into detail to discover something more than detail," and so this inquiry sets forth with the aid of clues suggested by the paradoxes that animate it.[54] First, the puzzling disconnect between broad popular acceptance of the case for active living and the pervasive underproduction of physical activity argues for investigation of production functions that are shaped by confrontations and accommodations—between the traditional metropolitan model of the built environment (with its mostly overlooked connections between built environments and active living) and the revisionist, active-living-friendly challenges mounted by smart growth and kindred persuasions—that unfold amid meandering efforts at "sense-making" in the cultural contexts of particular communities.

Second, the presence of active living on but never atop the agendas of the complex formal organizations that have (or could have) a hand in its production calls attention to how such plans and projects are mediated (and sometimes derailed) by a variegated cast of organizational actors in the private, voluntary, and public sectors and at all levels of the U.S. federal system. And third, the difficulties that dog the realization of active-living-friendly strategies that seem to be technically straightforward highlights how the (largely contingent) emergence of political champions in reform-minded localities, although necessary for the mobilization and assertion of countervailing power, appears not to be in itself sufficient. How crucial political championship can be both secured and sufficiently supplemented is a question that pervades the inquiry. These three analytical foci shelter a sizable store of variations within and across communities seeking to promote active living. Whether those variations also sort into themes remains to be seen.

1

WILKES-BARRE

Active Living on the Trail to Recovery

Wilkes-Barre, a city of 41,000 residents and the county seat within Luzerne County (population 320,000) in northeastern Pennsylvania, prospered in the nineteenth and first half of the twentieth centuries amid rich anthracite coalfields, the extracts of which were shipped to market in Philadelphia and beyond via a network of railroad lines and canals. From a peak of 86,626 in 1930, the city's population began a steady descent as the coal industry declined, and Wilkes-Barre fell on hard times. Economic activity dwindled, as did the supply of jobs, the retail occupancy rate, the tax base, and the appeals of the increasingly shabby downtown to better-off residents, many of whom decamped to surrounding communities, leaving a poorer, older, and less healthy inner-city population. In 1972 flooding from Tropical Storm Agnes, which left the city under nine feet of water, added natural insult to economic injury, and a recuperative tilt at federally funded urban renewal, an interviewee recalled, served mainly to "tear down good housing stock and replace it with public housing, which further depleted the tax base—and with office buildings filled with banks, which further accelerated flight to the suburbs."

At the turn of the twenty-first century Wilkes-Barre's private-sector power structure was, on some counts, poorly equipped to rebuild its economy. In contrast to Pittsburgh, a source explained, the coal barons of yore had left little civic legacy (such as local foundations) to the community, and headquarter firms had been replaced by branch plants. "It makes all the difference," the source continued, "when you need dynamic leaders that can make decisions, there used to be eight or ten guys who could sit in a room and raise money for X or Y. Now you would need 500 of those guys and you can't get them to do it."

However, the city enjoyed a distinctive and imposing power center of its own—namely, the Diamond City Partnership, a robust nonprofit downtown management organization that emerged from the work of what one business leader described as an "untraditional" Chamber of Commerce that acted as the regional economic development organization for the Greater Wilkes-Barre area and in that capacity took the lead on real estate projects, industrial financing, and plans for economic development in the region. Launched in 2001, the not-for-profit partnership was governed by a forty-member board and raised its budget by a combination of assessments on its members and contributions. Within the organization a cadre of energetic downtown business leaders elaborated and lobbied for plans to restore the city's growth and prosperity. The municipal elections of 2004 brought them an enthusiastic partner, a new mayor who attributed the city's woes to something deeper than the fiscal depletion that left it hard-pressed to maintain basic municipal services—namely, a civic culture that had lost its self-confidence, a "negative mentality within the mindset of the city" that far surpassed mere bricks-and-mortar matters in shaping the city's fate.[1]

The private and public partners asked themselves a simple question—among the many cities in and around the

Wyoming Valley of northeastern Pennsylvania that competed for business, jobs, consumers, and middle-class residents, what was Wilkes-Barre's competitive advantage? They came up with two answers. First, the downtown, properly developed, could become a "destination" for people seeking a walkable urban core in which they could not only work but also shop, dine, see a movie or play, admire "spectacular architecture and beautiful riverfront parks," study (a college and a university housed an estimated 7,100 students downtown), and reside.[2] Second, the city is surrounded by a natural environment of parks and trails that invited walking, biking, and other forms of recreation that would appeal not only to denizens of Luzerne County but also to ecotourists eager to follow in the great outdoors the historic journey of coal from mine to market. Indeed, with the right planning and funding, some trails could be connected directly to downtown venues—for example, a walking area built atop levees erected to prevent devastation from future floods.

Fixing the downtown was necessary (and perhaps sufficient) for a revival of Wilkes-Barre's fortunes: if the center languished, the city would never recover. The recreational and aesthetic pleasures of the city's natural environs were strong supplementary selling points. In short, although the term "active living" rarely appeared in the rhetoric of the downtown-enhancing businessmen and trail-expanding environmentalists determined to make Wilkes-Barre a regional—indeed, national—"destination," these leaders, like Molière's bourgeois gentleman, had adopted the prose of active living all along.

ACTIVE LIVING BY DESIGN

When the Robert Wood Johnson Foundation solicited applications for its Active Living by Design (ALbD) program in 2002, leaders across the business, voluntary, and public sectors in Wilkes-Barre, encouraged by a community planner in the National Park Service, saw in the grant a chance to deliver new funds, staff, coordination, and visibility to a crucial link in the local logic of city renewal—namely, "promoting and developing a trail network across suburban, rural, and urban communities" while undergirding that logic with an extra appeal: better health.[3] The planners had their work cut out for them: an ALbD activist noted that Wilkes-Barre was "a very difficult community, where one out of four people smoke, obesity is out of control, and public and private resources are impoverished and diminished." A business leader depicted the city as a "a prototypical rust belt population—work hard, party hard." A community activist added that the citizenry had "good intentions, but they like what they eat and are happy to sit on the couch. It's hard to get them to change pleasurable behavior. Everybody knows it [active living] should happen, but it's hard to make it happen. And the health messaging community isn't clear about what works. Do you scare them, urge them, or what?"

The project, which was housed at the local office of the Maternal and Family Health Services and proceeded in close collaboration with the Wyoming Valley Wellness Trails Partnership, had neither the money nor the leverage to alter or advance the renovation of downtown or the construction and connection of trails, so its leaders perforce focused on programming and promotion—the demand side of the active living equation. This they did in five main ways. First, the ALbD activists reiterated how the health "angle" enriched the rationale for changes in the built environment that were planned and under way—how, as a business leader put it, "active living and community development are a 'win-win.'" A colleague seconded this sentiment: "Community development means a healthy urban community—the principles are the same. We're fellow travelers, but framing does matter. Walkability in terms of public health gets attention that 'good urban design' might not."

Second, as an advocate explained, across the region's thirty-six municipalities and amid their abundant wellness committees, the grant enabled the project head to "maintain a focused presence of people on the ground who could spend time pushing active living issues forward and to provide credibility and consistency for a lot of fragmented things going on with little visibility and little collaboration. The grant gave us a leader and a focal point to pull things together and to build dialogue among different sectors of the community." This effort sometimes bore valuable fruit. In 2010 a trail activist contended, for instance, that the "partnering and connecting by the grant" partly explained why the trail groups succeeded in getting a local trail designated as a National Recreation Trail.

Third, ALbD brought an important constituency—the "health people"—more closely into conversations with business leaders, trail promoters, engineers, and others who presided over plans for the built environment. Wilkes-Barre's health institutions (the largest source of employment in the city) were not absent from the push for active living (for example, Blue Cross and Blue Shield sponsored a weight loss campaign between 2005 and 2007 and joined with the Chamber of Commerce to sponsor Healthy Workplace awards) but their involvement was, on the whole, thin and sporadic. Some payers and providers had responded to state cuts in Medicaid and other budgetary pressures by downsizing or eliminating their community relations staffs and their efforts at outreach on behalf of health promotion.

No health care institution could display public indifference to the virtues of active living, but some observers concluded that their hearts were not in it. As a county official put it, "these systems are paid for visits, treatments, and medications. There are good people in these systems but their leaders don't see enough [return on investment] from prevention." A physician who worked for a health insurer conceded that "maybe active living doesn't get the priority it should have, but the core business comes first, and that's the medical services, breakthroughs, and such. People in this organization 'care' about trails, but it's just not sexy enough." Moreover, the occasional exceptions (for instance, an ALbD activist noted that one local system had invested in a healthy communities program in nearby Scranton to the tune of $1 million per year) did not invariably meet collaborative expectations. "The health systems," a source observed, "want to own whatever they work on. They're willing to convene, but not to partner."

Fourth, the ALbD leaders supplemented and sought to extend the efforts of voluntary organizations such as the region's YMCAs and various disease-based coalitions to augment and publicize the impressive portfolio of active living enhancing events that transpired in and around Wilkes-Barre. These occasions included walking groups, festivals along the trails, hike/bike events, the Keystone Active Zone Passport project (intended to "get people to use trails and parks that are close to home for 'fun'—you're better off not saying 'health'"), and geocaching ("bury a treasure, like a piece of slate with a fossil imprint, in the woods or wherever, then they search for it with a cheap GPS that'll get you within 10 feet of it. Some people do 40–50 of these a year, and its real physical activity!"). Correlatively, the ALbD activists assiduously (and successfully) lobbied the local newspapers and TV and radio stations to inform the public in prime time about these happenings.[4]

Fifth, they tried to quantify the effects of publicity and promotion on the uptake of such activities by stationing counters along trails and at strategic sites to record the number of walkers and bicyclists who came by.[5] The findings were of more than academic interest: "We have real continuous measures of use, and can show a 40 percent increase from one year to the next on a new trail," explained an ALbD leader, who contended that the "measurement stuff" was the grant's main accomplishment. A trail activist added that the quantification "helps us make the case to county funders that the trails are really being used." (On the other hand, three rounds of counting in the downtown found little increase or decrease in foot traffic.)

As the grant wound down in 2007, one could say little more than that it had—probably—marginally elevated active living as a means to better health on the agendas of local actors. Observers could fairly describe the glass of outcomes as half full or half empty, as they pleased. The presence of a full-time coordinator had increased the volume of convening, conversing, publicizing, and programming around active living in Wilkes-Barre. A panoply of projects and events had attracted turnouts that some onlookers found impressive and others, disappointing. The persuasive skills of the project's director had brought more "health people" to more meetings and had made active living more prominent in their conversations, but the costs of turning talk into action rarely made business sense to the health professionals.

Nor had all the demand-increasing energy done much to build cultural and political support for an active-living-friendly built environment. "My job is to get people outside," declared one ALbD leader:

> OK—now bring in the built environment and we lose them. That stuff has to be translated into people-speak. The fact is, not that many local players "get" the built environment and its importance. Mainly, "active living" and "health promotion" mean telling people to eat better and exercise more. But active living through

> the built environment is another universe of analysis and discourse. When the selling and marketing of active living isn't just about exercise but also about the built environment it gets more subtle. There are planning connotations, with collective and community responsibilities, not just individual responsibility.

Moreover, the innovative nature of the counting exercise ironically limited its utility ipso facto—other cities of similar size seldom amassed similar documentation, so who could know whether the numbers and trends captured in Wilkes-Barre were impressive or marginal?

One estimate of daily pedestrian volume in a section of the downtown counted 7,572 walkers per day. By contrast, a comparison site in Chapel Hill, North Carolina, saw an estimated 14,184 pedestrians per day. The 1100 and 1700 blocks of Walnut Street in downtown Philadelphia registered 14,158 and 26,903 walkers per day, respectively.[6] And, counting aside, the effects of the grant program per se in shaping the demand for active living in and around the city were impossible to disentangle from the many other forces in play in Wilkes-Barre's challenging milieu. As will emerge later in this chapter, the main achievement of the grant lay in the nebulous (but crucial) realm of cultivation of human capital—in this case, the political socialization of active living proponents who came to deploy their commitment and analytical talents in institutional venues that had a hand in shaping the built environment.

DOWNTURN AND DOWNTOWN

As the ALbD grant phased out, the Great Recession of 2008 descended on Wilkes-Barre's beleaguered economy and left local officials, as a business advocate lamented, "just trying to keep the lights on." Budgets were tightly constrained in Luzerne County (which fired its wellness director and reneged on funds promised for programming in the riverfront development in Wilkes-Barre's downtown) and in Harrisburg, which (as a local official explained) left the city's Department of Public Works with transportation funds only for "repair and replacement but not for capacity-adding" projects. Cuts in agency staff and funds, driven by reductions in federal and state aid and by the mayor's efforts to pay off the city's heavy debts and to restore its credit rating, meant that Wilkes-Barre's Department of Health (one of only a handful in Pennsylvania, which addressed public health mainly by means of state district offices) was hard-pressed to fulfill its core duties let alone to innovate on behalf of active living, which consequently stood, as a source put it, " on the periphery" of its tasks. But the downturn did not dim and, indeed, served only to reaffirm, the commitment of the mayor and downtown business leaders to renew the city's inner core, an aspiration to the achievement of which walkability happened to be central.

Walking

The connection between economic vitality and the appeals of walking in Wilkes-Barre's downtown built environment was straightforward. Walkability enjoyed a sizable natural constituency among mostly car-free students, plus the faculty and staff of Wilkes University and Kings College, housed downtown. New apartments and condominiums were renting and selling well. A Barnes and Noble, a Boscov's department store, restaurants, a multiplex movie theater, a hotel, and an avidly awaited grocery store ("finally there are enough residents to support it") added impressively to the downtown's stock of venues to and among which pedestrians might want to stroll. Many of course would commence walking after driving downtown and parking their cars, so new infrastructure (for instance, garages and the creation of new on-street parking by moving buses and taxis to a new Intermodal Transportation terminal) was essential, and no less so was brighter and more plentiful lighting on the streets, which enhanced the sense of safety among those who used them.

Streets that were "calmer" ("the idea," explained one activist, is to create "an environment in which cars are forced to recognize that you shouldn't speed here") and more appealing to walkers in and around the downtown further helped to frame the picture. As a business leader explained:

> The opportunity for recreational walking is only part of building a culture of a healthier lifestyle. Parks aren't enough. People will walk only if it's a convenient way to live one's life, so a walkable downtown is crucial. We're trying to position downtown as a mixed-use environment and a 'walk to everything' site. Pedestrian access to so many things downtown is key to the appeal. The latest generation of the workforce, the millennials, want urban living, and we believe we can be the best offer of that in Northeastern Pennsylvania. Walkability is our competitive advantage as a residence and as a place to go to school. The suburbs can't compete. We can be a destination for the whole region.

Although business was, to be sure, the prime mover in the quest for active living downtown, it found support among several nonprofit (and tax exempt) partners who, for example, hired extra police and helped to keep the sidewalks clean. Wilkes University acquired state funds for traffic calming, pedestrian improvements, and better lighting on "its" streets and was working on a walkway that would connect the core of its campus with South Main Street. And an association of downtown residents was said to be "very active" in voicing its preferences to the Department of Public Works, the city council, and city hall.

These municipal improvements did not always meet with applause. A downtown business owner complained, for example, that although the new transportation terminal freed up new on-street parking spaces, it also reduced foot traffic by bus users in Public Square.[7] Another business leader wryly recalled how the largely state-funded redesign of busy River Street, proposed "10–15 years ago, was tied up by opposition for six years, and is now [2015] about done."[8] Locals had protested that the configuration, which installed extra pedestrian crossings, added new medians, and calmed traffic on the street by reducing parts of it from four lanes to two, would displace traffic onto adjacent streets, slow access to the local hospital, complicate life for emergency responders (a concern that led the planners to widen the lanes) and, as one memorable brick-bat had it, "privilege people over cars."

Meeting demands was, moreover, often as challenging as meeting objections. Mid-block crosswalks were a case in point: "They're popular. We have long blocks here and it's a big pedestrian issue. So we go to PennDot [the Pennsylvania Department of Transportation] and they say: 'That's what corners are for.' Mid-block crossings are chancy, 'cause people don't look, which means liability issues."

Opposition and obstacles notwithstanding, by 2010 local leaders were cautiously confident that they had learned how to manipulate the multiple public and private levers that determined whether their image of a walkable downtown would be realized. To illustrate, a business leader contrasted the River Street project with an earlier misbegotten effort:

> With Coal Street, people didn't focus on it till it was too late, and you gotta pick your battles. We're catching River Street at the beginning. The city engineers, PennDot, and so on would be offended if we said they're not taking care of pedestrians on Coal Street. They're going by the manual! Without relentless emphasis by key stakeholders to change the design, the status quo takes over. They'll build sidewalks—but without a push they won't care about trees and buffers and the like. On River Street we may see a better result. The Division of Flood Authority, the Chamber of Commerce, and other voices have said you've *got to* do traffic calming or people can't walk to the park. The state—PennDot—is in charge, but Harrisburg is one thing, and the guys in the district office who supervise the contractors who do it, that's another. You gotta be able to navigate channels of the bureaucracy.

By (and after) 2010 the relentless emphasis and eternal vigilance of the proponents of walkability seemed to be paying off. One proponent conceded the baleful impact of the Great Recession but underscored the "great strides" afoot:

> Pedestrian vibrancy has increased dramatically in the last three years. Yes, credit market troubles have slowed retail starts here, but new ones do open. Downtown has become an entertainment destination now. Before, people would say "let's go downtown to X." Now it's "let's go downtown." There are multiple destinations now, and they'll walk to them.

Biking

Biking, on the other hand, lacked the strategic salience that walking enjoyed. The downtown housed a free bike-share operation at a downtown hotel, but absent sustained advocacy by what one observer called "a couple of quiet bike groups," bike lanes remained at the farther margins of local policy. A public works official recalled that advocates who had proposed a bike lane years ago had been told by the traffic committee of the city council to conduct research on liability issues and costs and to get a professional engineer to propose a design—and had not been heard from since. The issue naturally arose in the redesign of River Street, noted above, but bike lanes entailed unacceptable tradeoffs. For instance, the traffic calming that reduced the number of road lanes was said to leave too little space for a bike lane, especially given the need for convenient on-street parking that would encourage people to come to walk in the park.

Bike leaders identified two spacious, busy, and central streets as good candidates for lanes, but the proposal made little headway in the city's planning office (shrunk by staff cuts from four officials to one, a transportation engineer, who was said to dismiss the lanes as a figment of "that weird health project") and the proponents were advised to enfold lanes into a "comprehensive plan" for integrating them into downtown traffic patterns. Besides, the city already offered opportunities for biking, noted Mayor Thomas Leighton, who cited as evidence a twenty-mile bike trip that he and his wife had recently made in and around the town.

The abundant trail projects in progress around, and sometimes within, Wilkes-Barre seemed prima facie to offer a built environment more conducive to biking than did the congested downtown. In this case too, however, hurdles were high and energies and resources with which to surmount them were limited. After joking that "kids ride bikes, not adults," a trail leader remarked that on a proposed rail-to-trail that would link with the downtown, "there is room for a bike lane, but it would be an unmarked one, and it's not in the budget, and it is on a state road. A company here will paint the symbols, but first we gotta get PennDot's approval."

Neighborhoods

Residents of Wilkes-Barre were occasionally heard to mutter that city officials worried much more about the revival of downtown than about the condition of the rest of the city and to greet pronouncements about improvements in the downtown with sarcasm—"Oh, the downtown is walkable so everything must be OK!" One critic contended that the economic importance of the downtown and the vigor of the association of downtown residents gave the city government an excuse to "fiddle while Rome burns" and to let the weakly organized neighborhoods "cave in."

The mayor, however, pointed to a list of the improvements over which he presided outside the core—fixing collapsed sewer lines, repairing damage done by floods in 2011, paving streets, rebuilding catch basins, demolishing dilapidated houses and turning the land over to people prepared to build new ones—and his reelection in 2008 and again in 2012 seemed to confirm that he was perceived to be doing all he could within the limits of strained municipal resources. Moreover, as one active living proponent observed, the neighborhoods were in some ways well suited to encourage walking: "Wilkes-Barre has lots of sidewalks and they are in good shape. The city is pretty good at resurfacing and so on. And lots of Wilkes-Barre people don't have cars."

The big deterrent to walking (and biking) in the neighborhoods, interviewees concurred, was by no means unique to

Wilkes-Barre—namely, safety. As elsewhere, crime sowed fear at both the macro level (the city had an unprecedented thirteen murders in 2013) and the micro (for instance, the much-discussed tragedy of a forty-nine-year-old "neighborhood watchman" shot dead in his front yard as retaliation for having called the police). In an article titled "Gun Violence Shoots Up 21 percent in Wilkes-Barre," the local newspaper helpfully observed that "state police suggest it is safer to reside anywhere else in Luzerne County" than in the city.[9] (In similar vein, the entry on Wilkes-Barre in Wikipedia [December 5, 2014] proclaims that the city "has a higher murder rate than Detroit and New York City," and that "Drug and related crimes are at an all time high with little sign of slowing down.")

Noting that the city is traversed by Interstate 81, the "heroin highway," which bears drugs back and forth from New York City and other venues, a city official attributed the rising crime rate largely to an ironic consequence of more vigorous law enforcement: pressure on gangs intensified competition among drug dealers and thereby generated more assaults and murders. In fact, the murder rate had fallen since 2013, the official contended, but he also grumbled that this consoling datum got little attention in a media climate in which "if it bleeds, it leads."

No one disagreed, however, that perceptions of crime discouraged active living in the city's neighborhoods. Certainly, this applied to much of the city's Medicaid population, as a former outreach worker at a health plan remarked: "You can't just say go out and do exercise. Safety is uppermost for them, so you've gotta look at that environment and work with it realistically. For instance, are there safe connecting walkways? No one really cares for these folks, though. It's a silent, forgotten population with little political influence." Parents who were terrified that their kids might encounter strangers, an ALbD leader observed, withheld participation in safe routes to school projects, and kept their kids "disconnected from nature. Some won't even let them play in their own yards after school till mom gets home." The aged "are afraid to walk outside for fear they'll step on needles," lamented a source.

Nor did fear of encounters with shady characters and their transactions end upon entry into the protective precincts of the downtown. Officials hoped that a combination of bright lighting and easy parking on downtown streets would calm these anxieties, but the impact, both immediate and potential, of crime on their plans for renewal caused anxieties of its own. As one business leader put it:

> Thirteen murders in 2013! That's unheard of in Wilkes-Barre. It's poisoned everything in the public discourse. "Downtown is dangerous! You need a flack jacket to enter Public Square!" People say: "I gotta get out of the city, sell my house—which *is* walkable—and go to the suburbs where people can't walk to *your* house." And the social service agencies are downtown, so drug types go there. There's no drop-in center, so they use the library! We could fix it, but the social service agencies have no money. It's hard to get people to use River Street Common if they see people shooting up there. It really does matter. We have massive perceptual issues with crime, and it's the number one thing we have to address.

The top five weaknesses of the downtown, according to a survey conducted in 2014 by the Diamond City Partnership were "Feels unsafe," "Impact of Loitering and Uncivil Behaviors," "Impact of Concentration of Social Services," "Condition of Public Square," and "Impact of Homeless Population."[10]

TRAILS IN TOWN AND COUNTRY

Active living benefited collaterally not only from the city's work to revive its downtown and to stabilize its neighborhoods but also from its diligent efforts to build and expand trails around the city and to connect them to it. The expectations that Wilkes-Barre's leaders entertained for trails were precise and ambitious. As an enthusiastic business leader explained in 2007, the city

> has been involved in planning trails development for 30 years and has been executing those plans for 15 years. By connecting core trails, raising the riverfront levees used for flood control, and using railroad rights of way, we want to get 150 miles of connected trails. It tells part of American history, like they do in Lowell [Massachusetts]. It's the coal corridor, from mine to market for home heating. It starts in downtown Wilkes-Barre and involves five counties. And it's almost done—one link across a mountain, that's the last part, and then you could walk and bike it! Likewise with the levees. Don't just build a wall against floods. One local engineer here singlehandedly convinced skeptics at the Army Corps of Engineers that the levees were a recreational opportunity. When it's done there will be 13 miles of levee-top trails you can walk or bike on. And we're using parallel rights of way on anthracite scenic trails for walking and bike paths. This is another six miles, and then you could walk or bike to work in the center of Wilkes-Barre. Anyway, that's the vision: a metro trail system connecting the Greater Wyoming Valley communities.

The "metro trail system" sketched above evolved from the interplay of myriad organizations—mainly not-for-profit and driven by an intense combination of what J. Q. Wilson called "solidary" and "purposive" incentives[11]—that had a piece of the proverbial action in pursuit of goals that embraced aesthetic improvement, environmental protection, expansion of ecotourism, historic preservation, reclamation of mine-scarred land, regional economic growth and promotion of active living. Because planning, constructing, extending, improving, and connecting trails require capital, these organizations perpetually canvassed a range of funding sources, primarily public, at the federal, state, county, and (occasionally) local levels of government. The most prominent in the cast of institutional characters were the federal Department of Transportation, the Environmental Protection Agency, and the Federal Emergency Management Agency (for repair of damage caused by floods); the Pennsylvania Department of Transportation (PennDot), Department of Conservation and Natural Resources (DCNR), and Environmental Protection; and the Recreational Facilities Advisory Board in Luzerne County.

Trails advanced when the demands of entrepreneurs for specific projects (which could emanate from individual groups but often were promoted by coalitions and partnerships among them) aligned with the missions and agendas of suppliers of funds. A trail-builder illustrated the process by sketching the efforts of a regional greenways partnership to spur trail and river projects:

> Each lead organization in the partnership proposed projects. For instance, in one town where the river runs near historic buildings, there was an opportunity for boating, a camp site, a pavilion, and rest rooms. The projects incorporate the river, history, and natural conservation. We partner with the local conservation district—for instance, there is a bald eagle nest near the town.

> And they incorporate economic history. The coal industry scarred this landscape and our organizations are reclaiming it. We corral the agencies and funding streams, agree on concepts for projects, and sell the idea, emphasizing what's in it for them. You need that broad-based appeal or the project won't work.

"What's in it for them," meant little, however, if "they" lacked the *quid* to exchange for an appealing *quo*, and federal transportation monies had become more elusive over time. In 2015 a trail leader lamented that these funds "were more threatened with every transportation bill" in Congress. Securing federal monies, moreover, required navigating arcane and insular channels between the federal transportation authorities and PennDot, which also found its own resources increasingly constricted. Funding by the state DCNR and by Luzerne County entailed less formidable bureaucratic hurdles, but sums were smaller—and declining as well. In 2010 a trail leader deplored a "terrible" financial situation in which the number of grant applications funded by Luzerne County and the state DCNR had plummeted, as had also the size of the (few) awards.

Constrained fiscal environments increased the need to mobilize political champions, who sometimes resided in the state bureaucracy. A business leader averred:

> Without the DCNR, trail development as we've done it here would have been impossible. For 15 years people at the highest levels in DCNR have emphasized riverfront development and trails. Key people have been plugging away in Harrisburg for years, and that gives us people in the right places at the right times to make the right local connections. We could pick up the phone and call these state people. They were experienced and knew how to use the state machinery.

Such connections were necessary but, in the face of constrained budgets, not always sufficient.

At times, elected officials befriended the trail builders in their search for funds. A leader recounted, for example, how, when an organization that had promised $1 million to build a trail bridge over a railroad track reneged, "We spoke to our local congressman, who is a native of this area, and our state reps, and they resolved it."

Beyond the securing of funds for trail projects lay many strategic hurdles.[12] The most important include securing ownership of, or rights to passage through, land that is essential to the completion or connection of trails but is owned by a private party or public agency—a task that can take years if the property changes hands in mid-negotiation or if the owners are recalcitrant; obtaining insurance against liability in case someone is injured by a fall or other mishap while using a trail (describing his group as "a 501(C)3 that has to beg for money," a trail leader explained how his volunteers had held a pizza sale to raise $2,000 for insurance); keeping trails and their users safe from troublemakers; providing adequate parking for those who need a car to get to the trails (which are a fair distance from many residential areas); and managing tradeoffs between scarce resources. Are exercise stations along the trail a good investment or a waste of money when compared to, say, gnat eradication? Should trails be paved—which, to some aficionados, makes them less "natural"—and, if so, how far and with what surfacing? A costly grade of crushed limestone makes life easier for bikers and for hikers in wheelchairs but reduces the amount of surface that can be paved or repaved.

Once trails were in place, what funds and personnel would maintain them? A member of a statewide nonprofit organization sketched the problematic division of labor:

> We can do a lot, but it's better if the municipality is involved. We gotta convince them that the benefits are greater than the costs. The carrot is they get first-dollar money for planning, design, and building the trails. But then management and maintenance are all theirs. And the municipal involvement is not so good. Planning these projects takes 12 years, maybe, and the municipal leaders say "come and talk to us about maintenance when it's gonna happen." And it does take so darn long—so now we have several trails that are just opening, and there's been all this turnover in the municipalities, so management and maintenance is a continuing challenge.

Meeting the challenge meant mobilizing volunteers to keep the trails in shape and to repair them in the aftermath of floods, storms, and other sources of erosion. Help could be obtained ad hoc from groups such as the Trail Tenders and Vista Volunteers, but steady support for core personnel remained elusive. The trail organizations, said one member, constitute "a very small group of diehards," mainly seniors. The difficulties of relying on "the same folks, over and over," a county official added, argued for entrusting maintenance to some sort of public–private partnership, a proposal that had long languished on the lower rungs of the county agenda.

Trail development in and around Wilkes-Barre proceeded with the health care sector largely on the sidelines. A physician-manager at a health plan acknowledged that trails afford "the easiest type of exercise," and his organization, like others in the area, was pleased to publicize to both employees and the general citizenry upcoming hike/bike events. But physicians were said to be leery of writing "exercise prescriptions" for the use of trails (would they be sued if someone got injured while following doctor's orders?), and the engagement of payers and providers went little beyond intermittent attendance at meetings of the plentiful coalitions and committees that addressed "wellness" (and perhaps the development of trails).

The Wyoming Valley Wellness Trails Partnership worked to bring the health people into closer contact with the trail people and, as noted above, became in turn the major conduit between the two camps for the ALbD project in the 2000s. The new synergy seemed to make sense. On the one hand, a lack of resources and authority obliged the active living advocates to concentrate on the demand side of the built environment—getting people out and about on the region's impressive network of trails. On the other hand, the advocates perceived that the trail groups lacked both resources and zeal for promotional endeavors. As a frustrated activist explained,

> There are three levels to this—acquisition of land and construction, then operation and maintenance, and then promotion. The first job is to assemble the money. The second is not so easy. And some of the trail folks want any "extra" money to go to operation and maintenance, not promotion. Marketing just isn't their thing.

The trails–health partnership promised to make winners of both sides: more promotion meant more use; more use, properly documented by means of counters, meant more evidence of demand; more evidence of demand meant success for the ALbD project and ammunition for the trail groups in their search for funds. Trail leaders were confident that the expansion of partners led in turn to the enlargement of participation in trail-based activities. One mapped the complementarities in 2007:

> Local industries are getting more helpful. Companies that are getting into wellness now are calling us. At one hike-bike event,

> the public was invited, we had picnics and so on. One company provided 500 hamburgers—and there were also healthy alternatives like carrots! PennDot did bike fittings and gave out helmets, free while they lasted. And the Active Living program had ponchos for the first 30 kids who showed up. We publicized it through the two TV stations, on the radio, in the paper—a community awareness thing. People came from all over. One hundred fifty to two hundred people turned out—we get more each year.

Notwithstanding the absence of detailed baseline data, early returns from counters funded by the ALbD grant did indeed suggest that the trails were being used more frequently. For example, speaking in 2010 of a "four-mile trail without connectivity yet," a trail leader cited an increase of eight thousand visits between 2008 and 2009. As is usually the case in the active living domain, the explanations postulated for this surprising surge were overdetermined (i.e., anyone's guess): word of mouth; local publicity deriving from the joint efforts of the trail groups and the Wyoming Valley Wellness Trails Partnership; the ALbD money the partnership received to sustain the publicity; and perhaps the appeal of the once-controversial exercise stations. Yet another potential draw, soon about to launch, was "an environmental education area, with a garden, benches, a butterfly area, and information about warm season grasses and so on, that lets students learn from what they see out on the trail. This had been an unsightly area—no crab apples and red maples! But we got 20 trees from the Forest Service under the stimulus package!"

To be sure, some observers identified lacunae: bikers made little use of the trails and lower-income, inner-city residents seemed to be immune to their appeal. All the same, interviewees voiced cautious optimism about the future of the rapprochement between active living and the trails enterprise. Funders increasingly demanded evidence of partnerships among their applicants, and the addition of the health angle and its organizational embodiments could be helpful. Employers worried about days lost from work by unhealthy employees, a problem for which publicizing the advantages of exercise on the trails was an inexpensive and uncontroversial corrective. Schools had begun measuring the body mass index of their students and felt compelled to "do something" about troubling levels of obesity. In sum, the trail groups had found in the health sector prominent (albeit aloof and intermittent) collaborators, and the advocates who sought to promote health by means of active living were awarding trails a higher priority in their strategic portfolio.

PARKS AND RECREATION

Parks are a conspicuous glory of Luzerne County. In the words of a county official: "There's a perception that this is a strip mined, burned out area, but you can be in ten different parks within ten minutes of here!" Set both in and around the city of Wilkes-Barre, and sometimes connected directly to trails, the parks offered residents and tourists attractive sites in which to combine leisure with physical activity.

Parks do not, however, build, maintain, and repair themselves, an inconvenient truth that triggers thorny questions about the division of responsibility for them among levels of government. Luzerne County had its Department of Parks and Recreation, and so too did the city of Wilkes-Barre—not unusual in Pennsylvania's universe of small, incorporated localities, each with its own mayor, police force, and other municipal bodies—a form of fragmentation deplored by critics who had long argued, unsuccessfully, for regional governance of parks (among other things) in place of an endless series of ad hoc agreements among jurisdictions.

At both the county and city levels, parks (like trails, bike lanes, and new sidewalks) risk being marginalized in tough economic times as non-essential objects of public spending. In 2007, before the impending economic downturn, a county parks official lamented that his agency was viewed

> as fluff, an afterthought. For example, we have one beautiful park without winter programming for want of money. In this department there are four [full-time employees]. A ranger makes $22,000 a year. Our vehicles are antiquated and some have 150,000 miles on them. Our county leaders do value parks and recreational opportunities for kids and families, but there's a multi-million-dollar deficit. We get some funding through HUD Community Development grants, but there's no operation and maintenance money, so those parks are just fields now. Better to enhance existing ones, and for that we need volunteers, who come from struggling local not-for-profit groups. There are no deep pockets anyway—any we can use. The state has been helpful. Our local DCNR reps and elected officials are supportive of green space. But then there's no consistent zoning because the localism is so strong.

Three years later the situation had deteriorated, as a trails activist explained:

> The county eliminated the recreation department for 2010 and laid off a lot of workers. Our Wellness Trails members pleaded with the commissioners to keep them open. Soccer tournaments make a lot of money, for example! But recreation is

> a discretionary function, not mandated by the county code. The commissioners did raise taxes, and the budget for recreation was only $300,000, but we failed. The county has been in financial trouble for 5–6 years now. The city of Wilkes-Barre has its own department but it has no capacity to do county stuff. And the state is no help because it too is in tough shape.

Like other elements of an active-living-friendly built environment, parks benefit from advocates and allies who defend their importance to the community and seek to fortify their budgets. Sometimes this advocacy involves mundane pushing and pulling within existing institutional frameworks—a strategy that has its hits (for instance, $12 million in tax credits in 2008 and $950,000 in 2009 to renovate Coal Street Park, secured by the joint efforts of Senator Bob Casey, Representative Paul Kanjorski, Mayor Thomas M. Leighton, and state and county officials[13]) and its misses (for example, the unsuccessful appeals to retain the county recreation department noted above).

Occasionally, however, new structures of governance create openings in which entrepreneurs may try not only to practice advocacy but also to institutionalize it. For instance, a referendum in 2012 brought Luzerne County a reorganized county council (formerly called a "commission") adorned with new advisory boards intended to encourage greater involvement of local citizens. Seizing the opening, two leading ALbD advocates, who had honed their political skills and established their bona fides in promoting the program's various projects, were appointed to the five-member recreational facilities advisory board of the council and used their positions to promote access to, and improvement within, parks and trails for active living.

One facet of their agenda was financial. For instance, they successfully recommended that $40,000 of state funds (derived from a conservation-targeted well-head tax on natural gas drilling) be provided to complete a missing part of a trail in the southern part of the Wyoming Valley and worked to persuade the council to expand both its budget for parks and recreation and its efforts to secure a larger share of that state tax on drilling. Another was reorganizational. When in June 2015 the state agreed to take over from Luzerne County a park that had been in decline for years—a development that was expected to bring $1 million in upgrades, including improved trails and public access that would be integrated into the regional trail effort—a board member who had supported the transfer "called the woman who had started the study for that project in 1973" and told her: "'It's happened.' She said it was one of the happiest days of her life—and to think it only took 40 years!"

A third, interorganizational dimension of their strategy targeted parks directly. One of the advocates explained in an interview:

> We got the Luzerne County Transit Authority to run a bus to the state park. It's the only thing of its kind in Pennsylvania—a regular bus right to the park. And now we've got it up to three runs a day! There's no open swimming pool in the city of Wilkes-Barre, so the [county] park is important, and the 45-minute ride along that route goes past public housing projects. It gives real access to a park and a pool. And the transit authority lets older people ride free and likewise kids under such and such a height.

And sometimes the interorganizational objective was diffusion of information about innovations:

> The CDC has these "success stories" [brief narratives compiled and disseminated by the center's Division of Community Health,

> with the aim of "Making Healthy Living Easier"] and we do some of these. So we showed that model to the Pennsylvania Environmental Council, and they're doing it! It makes for great handouts.

Finally, political networking goals complemented their strategic aims. As one of the advocates remarked: "The idea is to support the outdoors-positive people on the council, to urge people to get on these boards, get people in who care." In sum, the human capital nurtured by the ALbD grant was in due course generating political capital and, thus, championship within important policymaking venues.

Viewed from on high, Wilkes-Barre looks nothing like a poster child for active living. Its streets are not, nor are they about to become, "complete"; a connected network of bike lanes is not imminent; its pedestrians are neither vocal nor organized; and the health benefits of active living do not figure prominently on its civic agenda—or in its civic culture. Viewing Wilkes-Barre from on high, however, ensures that the city's practical engagement with active living will be misunderstood.

The civic (indeed, county and regional) agenda pursued by partners in the public, private, and voluntary sectors in Wilkes-Barre centered on the renaissance of a decayed downtown, the cultivation of beautiful natural resources (especially trails and parks) in and around the city, and the physical and perceptual connection of the former with the latter in ways that gave "holistic" appeal to the city and its larger region as a place in which to live, work, study, and tour. In this context, walkability and, more broadly, the presence of environments—"built" in the downtown and "natural" along the trails—that enabled and encouraged physical activity was not merely a latent opportunity awaiting the labors of active living advocates to be made manifest; the presence of these environments was also an intrinsic component of the "action logic" of urban renewal, twenty-first-century style, in and around the city.

The central challenges for active living proponents were to "read" this city, county, and regional context insightfully; to deploy local knowledge in the identification of points and paths of intervention that reinforced elements in that context that were congenial to active living; and to craft political stratagems to persuade the region's multisectoral partners that active-living-friendly initiatives—traffic calming, street lights, sidewalk improvements, extended trails, maintained parks, events of all sorts that featured physical activity, and more artful and abundant communication to residents about the opportunities for and advantages of all of the above—deserve a more prominent position on their agendas and a greater investment of their time and resources. So viewed, active living is a political exercise that rises or falls with the sagacity and capacity of its practitioners to "make sense" to local leaders and residents as they seek to interpret and improve the particular political economy that fate dealt them.

If blueprints extracted from planning texts and commandments proclaimed by public health or other external authorities are to be more than merely suggestive, they should take sensitive account of the local particularities of civic action, gain the allegiance of multiple custodians and users of the built (and natural) local environments, and win more than passing political support from local leaders—who, as the Wilkes-Barre case makes clear, may be found in the private and voluntary as well as the public sector. Uplifting as it may be to declare that health is a top priority (without it, one has nothing) and to push for health "in all policies," in Wilkes-Barre the levers for health promotion are too few and too weak, and other priorities (fighting crime, lowering unemployment, stabilizing the tax base, and so on) are too exigent to permit health promotion to stand distinctly near the top of the civic agenda. The best—indeed, only feasible—course for health advocates, then, is to identify points of convergence and solubility between their agendas and larger policy priorities and to improvise strategies accordingly. Such strategies are visible in the city's partnering sectors, and the results, taken in context (as of course they should be), are not unimpressive.

2

LOUISVILLE

The Politics of Piecemeal Progress

Louisville (population 620,118 in 2018) is the seat and center of Jefferson County (population 756,832) and the largest city in Kentucky. (Lexington, second on the list, is about half Louisville's size). In 2003 the city and county merged into a unified Louisville/Jefferson County Metropolitan Government, also known as "Louisville Metro." Founded in 1778 as a port along the Ohio River, Louisville grew steadily as a hub for shipping and cargo, a railroad nexus, and a center for production of appliances and cars. At the turn of the twenty-first century, Louisville, like Wilkes-Barre (discussed in the previous chapter), wrestled with the forces of decline that U.S. patterns of metropolitan development inflict on most center cities, but it did so with three advantages Wilkes-Barre lacked. These were (1) a larger and more diversified economy that included several major corporate headquarters (Taco Bell and Papa John's Pizza, for example), prominent health care institutions including Humana (an insurance giant), and several medical centers; (2) expanded public authority to tax and regulate, which followed the above-mentioned creation of a consolidated Louisville Metro government, headed by a powerful mayor; and (3) a government system that, although not swimming in cash and personnel, had not (like Wilkes-Barre's) been cut to the bone by retiring accumulated debt and trimming the ranks of municipal agencies. Whereas the dominant civic narrative in Wilkes-Barre sketched a battered but proud "contenda" rising from the canvas to resume the good fight, the narrative of Louisville featured the quest of Kentucky's premier urban center to retain its standing and enhance its advantages in relation to regional and national peers in the eternal competition to attract taxpayers, businesses, nonprofit institutions, tourists, and favorable national attention.

The two cities also offered contrasting contexts for active living. In Wilkes-Barre, smart growth, New Urbanism, and active living were, in essence, luxuries the city could only afford to honor coincidentally (for example, a walkable downtown was an economically healthy downtown). Louisville had housed since the 1980s a cadre of professionals in the public, private, and voluntary sectors who championed these progressive movements and the active living agenda they encouraged. When the Robert Wood Johnson Foundation sought applications in 2002 for Active Living by Design (ALbD), the program found in Louisville a ready constituency that set about identifying the most promising "hook" on which to hang the grant amid projects then unfolding.

ALbD IN LOUISVILLE

When the request for proposals for ALbD appeared, planners in Louisville were laying plans for new public housing, supported by a Hope VI grant from the U.S. Department of Housing and Urban Development, in Liberty Green, a lower-income, largely African American community not far from the downtown.[1] Infusing active living into the designs for new housing seemed to be a target of opportunity for a "holistic" integration of better housing and better health. A walking corridor would connect the three principal neighborhoods within the scope of the Hope VI project—a scope that the city specified, as a planner explained, not because they were the neediest areas but rather for their proximity to downtown, the redevelopment of which topped the mayor's agenda. The corridor strategy promised at once to promote active living and to enlarge the domain of downtown revitalization in a package that reflected both New Urbanist perspectives on planning and design and those of an uncommonly broad cast of agency participants that kept the project from becoming (in the words of one source) "the usual introverted housing thing." As they drafted plans for streets, walkways, and parks that encouraged walking and cycling, however, the planners, working under the aegis of the city's Housing Authority, met unexpected resistance from the community itself.

One member of the ALbD inner circle recalled the reaction to proposals for a greenway that would connect sidewalks and streets in the neighborhood to a park.

> We had the right of way but we ran into conflicts within the community. There's the NIMBY [not in my backyard] factor: they don't want people walking around by their houses. Bike lanes can interfere with on-street parking. Older residents fear rapists on trails and hooligans on bikes. We learned a severe lesson from this: neighborhoods are not homogenous.

Many residents hesitated to get out and about for fear of encountering dangerous characters such as drug users and dealers and members of youth gangs. Asked about the value of eyes and people "on the street," an activist demurred: "Yes and no. It depends on whose eyes and which people." The physical improvements that in principle encouraged residents to walk and bike also increased the mobility of predators. "Enhancements" such as shade trees and streets becalmed of traffic gave troublemakers havens in which to watch and wait. Safety concerns trumped all others. And the health benefits of walking and biking sometimes failed to impress lower-income residents, who used these modes of transit only when they lacked cars and, an observer conjectured, aimed to own cars, as did the middle class, before they would consider (like the middle class) ditching the car for a bike or a walk.

Having hit unexpected cultural obstacles, the ALbD planners sought counsel from the Presbyterian Community Center (PCC), a prominent multiservice organization that was said to be understaffed, slow moving, and cumbersome but whose leader was widely characterized as the "gatekeeper" between city officials and the residents of Liberty Green. Staff at PCC conceded that the rigors of daily life in the Hope VI area discouraged active living: "We try to motivate physical activity, but it's not high on people's priority list. If you're working to feed your family and so on, it's not a high priority. The big issues are having a bed, not getting hurt, and staying out of jail." Girls posed a special challenge: "Once the boys hit middle school, basketball rules, and that can be quite a good workout. But there's maybe one team of girls. So many get pregnant; you can get your own housing unit if you have a kid." "Teenage girls are very much at risk," added another source, "but they don't want to sweat or mess up their hair, and they're very self-conscious. They'll dance but not 'exercise.' These are self-esteem issues that you gotta address."

Nonetheless PCC staff acknowledged that ALbD had made them more aware of the benefits of physical activity for people in the neighborhoods the center served, that exercise had risen on the center's agenda, and that ALbD had helped to forge closer relations between PCC and Bicycling for Louisville, the city's main organizational advocate for cyclists. But the center had no recipe for moving from theory to practice and could not afford a major promotion of active living of its own. "We offered a whole program of physical exercise—hip-hop-ersize and more," a staffer explained, "but with the economic problem the money dried up and those programs went away. We tried to keep it going with donated time, but we couldn't, and there's a freeze on our workout facility. It's no one's main focus."

As the salience of the safety issue grew clearer, the planners and PCC staff launched discussions with the Louisville police department. Relations between the community and the police were tense, however, and no one knew how to make active living in Liberty Green simultaneously healthy and safe. One modest gambit—assigning Housing Authority officers to patrol the neighborhood on foot—ended ironically after two weeks because the officers preferred to stay in their patrol cars.

Apparently promoting both safety and active living in Louisville's lower income neighborhoods entailed sharp tradeoffs. Interventions to promote physical activity arguably presuppose advances in such seemingly disconnected policy arenas as law enforcement, youth development, and after school programs. A police officer explained:

> We should develop safe bike routes, parks, playgrounds, lighting, and so on but it's not simple. Interior court yards are no good for playgrounds because the police can't see drug dealing there. Better to put them at the edges. People keep their shades down and are intimidated to go out and walk. There are dice games at the corners, and some people even call to complain about rowdy

> horseshoe players. People who won't walk through public housing might walk at the edges of it, but getting *to* the edges is a problem. And the old idea that if you close off streets in public housing you'll deter drug dealing is wrong. Absence of traffic encourages drug dealing because we can't get so easily to where they congregate. And some dealers go around on bikes! To get at dealers you should trim back the trees, remove the shady spots, and so forth. The biggest active living challenge is getting people *out*. In these areas, lots of people do walk and bike—from necessity. But beyond that, crime is a deterrent. We've got lots of gangs—Bloods, Taliban, and East End and West End Bloods fighting with each other! It's hard to keep kids engaged and out of the gangs. You have to have activities *and* keep tabs on them all the time. But moms aren't home—they're at work—and for very little money.

More modest and geographically concentrated projects were similarly stymied by local politics. An activist recounted that

> we wanted to do an active living center. We have $3.8 million for it but we can't move ahead. The city wants more remunerative uses for the land, some in the community want a gym, and middle-class types want a clubhouse. We can't get agreement. We pushed for out of doors activities—all that fresh air! The community balked: what if it rains, it's cold, forget it. They want indoor stuff. We've had to revise our ideas.

The impact of ALbD on the neighborhoods covered by the Hope VI grant was not negligible. A source conjectured that the sidewalks in the project area were more walkable and better connected than they might have been absent ALbD, and the program also drummed up interest in a school-based community garden and in farmers' markets, corner stores that sell fresh foods, and other enhancements that blend active living with healthy eating. The ALbD initiative, meanwhile, steadily morphed away from a neighborhood-centered strategy and into more diffuse campaigns that seized chances to promote biking and walking wherever windows of opportunity opened in Metro Louisville. With the assistance of a transition grant in 2008 from the Robert Wood Johnson Foundation (which supported a modest amount of continued staff time), the Active Louisville project became an active living committee within the Mayor's Healthy Hometown Movement, launched in 2004 and run by the Department of Housing (DOH). Fearing that this emplacement was less a new lease on life than a consignment to life support (or a bureaucratic burial), proponents of active living pondered anew how to concert and expand their influence within metro government and, most important, in the office of the mayor.

NEW ENABLING CONTEXTS

As the ALbD grant came and went, the policy landscape for active living, including both demand- and supply-side interventions, looked increasingly favorable. Like the other cities examined here, Louisville overflows with more or less "comprehensive" visions, plans, ordinances, codes, manuals, and regulations that purport to govern the development of the built environment, and in the first decade of the twenty-first century, these measures grew both more numerous and more profusely detailed in provisions that gave a higher priority, at least on paper, to the encouragement of walking, biking, and other types of physical activity.

In 2003, following the city–county consolidation that created Louisville Metro, planners pushed to include within a new comprehensive county plan "form-based" land development codes that sought to tailor rules to the "character" of the sites in question. (As a case in point, a planner cited Bardstown Road, which "changes its character and context from city to suburban to rural. So we ask 'how should the *road* change' and we try to develop form districts that reflect that.") Attention then turned to the promotion and adoption of Complete Streets—ones that provide accommodation "on all new and reconstructed roadways for ALL users."[2] This ensures that "people can safely get from point A to point B while getting exercise at the same time."[3]

In 2008, having won enactment of a Complete Streets ordinance, which promised to reverse what Mayor Jerry Abramson had called a decades-old "urban planning mistake"—to wit, the city "built roads only for vehicles"[4]—proponents of active living labored to refine their strategies to implement it. A planner explained:

> The major obstacles are educating policymakers, council members, and neighborhoods on the benefits of active living. Sometimes it's economics: a park is a nice place to have nearby, and it also boosts home values. How do you get developers to see this and to agree? You work with neighborhood planners, which we're doing now. You explain to the developers why we require x, y, or z. Some see it, but others are stuck in the 70s/80s mentality. Same with public officials. In Nashville, we learned, the planning director butted heads with the public works director on a highway project. The planner persuaded him that ten lanes would be OK instead of 12 or 13—because it's gonna be cheaper to maintain! Likewise with traffic circles—it slows down the traffic but it also adds home value. You gotta sell the economics.

By 2010 further signs and omens suggested that the advocates' sales pitches, economic and otherwise, were elevating the priority of active living on Louisville's civic agenda. New master plans for cyclists and pedestrians (the latter aiming to make Louisville "the world's safest and most appealing community for pedestrians") were adopted. Trial and error had led planners to discard strategies unacceptable to residents in the Hope VI neighborhood on which ALbD had focused and to embrace new approaches. By 2009 the Liberty Green housing development was said to feature "wide, unobstructed sidewalks, well-marked and signalized street crossings, way-finding signage, traffic-calming islands . . . and three pocket parks."[5] One observer, however, wondered sardonically how much of this progress derived from displacement rather than strategy: tearing down public housing units and dispersing their residents elsewhere while new buildings went up did indeed tend to reduce crime, and fear of crime, in the newly depopulated area.

New Hope VI housing projects were contemplating the corridor connections and other enhancements to walking with which the project at the heart of the ALbD grant had struggled. New authoritative sources (for instance, manuals issued by the National Association of City Transportation Officials, an organization formed in 1996 to promote "safe, sustainable, accessible, and equitable transportation choices"[6]) that challenged the biblical status of publications of the powerful National Association of Highway Engineers were more widely promulgated and increasingly familiar to highway engineers. In 2015 the city's Office of Sustainability issued Louisville's "first ever sustainability plan," which pledged allegiance both to the metro government's "health in all policies" approach and to alternative transportation.

In 2013 Mayor Greg Fischer unveiled Vision Louisville, a broad-ranging plan that sought to guide the city's development through 2040; goal number 4 exhorted the city to become a "Leader in Healthy and Active Living."[7] In 2016 the mayor presented Move Louisville, a one-hundred-page, twenty-year plan that took "a holistic approach to our [the Metro region's] transportation system."[8] Fischer explained that Louisville must "attract more talent . . . and we hear this over and over again, today's young local professionals are less interested in owning and using cars. . . . They want . . . good public transportation, more walkability, more bike lanes."[9] In short, several enabling contexts—cumulative and mutually reinforcing—had come to frame the policy processes that shaped the built environment in Louisville. Emboldened policy entrepreneurs in the active living camp pondered how to implement these promising formulations.

BIKING

Marking off a portion of a roadway as a lane reserved for cyclists is not inherently expensive (all one needs is paint), but such innovations involve acts of public management of public space that threaten to upset the settled expectations of private users of the space in question. To bike enthusiasts, the merits of, for example, "flipping one-way streets to two-way so as to allow bike lanes with two-foot buffers" (as one put it), may seem axiomatically obvious; to the general public not so much. Residents and businesses complain of lost parking spaces. Commuters grumble about greater congestion and slower travel times when roads must be shared with bicycles. Bike lanes compete for the time and attention of the Department of Public Works (DPW; Louisville has a transit authority but no department of transportation), which juggles demands for new sidewalks, repair of existing ones, filling potholes, and other priorities. To move the creation of bike lanes (and still more ambitious, a definite and steady expansion of the number of *miles* of lanes) to a prominent position on the civic agenda and within the institutions that implement it requires a firm, continuing political push.

In Louisville this push came from an energetic advocacy group of cyclists, biking enthusiasts within city agencies (most notably the DPW), and city hall. Bicycling for Louisville, a not-for-profit organization that advocates for "a safe and convenient bicycle network and for the rights of people on bikes in Louisville and Southern Indiana," has worked since 2004 to advance the interests of bikers on both the demand and supply sides of the picture.[10] Guided by the five "e's"(engineering, education, encouragement, enforcement, and evaluation and planning) recommended by the League of American Bicyclists as the "essential elements of a bicycle-friendly America" the group sought to build support for biking among residents, the media, and political leaders by sponsoring activities such as classes on safe cycling and on how to maintain and repair bikes, by producing and posting on YouTube a rap video touting the virtues of biking, and by launching proclamations and events such as Bike Everywhere Month and Bike to Work Day.[11] Increasing the number of Louisville cyclists, the number of bike lanes available to them, and the facilities and amenities that encourage the use of bikes (for instance, storage rooms in residential and commercial buildings and designated spaces for bikes on buses) of course stood high among the group's priorities.

Moreover, the city's DPW included an official who had since his youth been an avid, elite-level cyclist and who, having acquired a master of public health (MPH) degree, had pressed the case for the health (and other) benefits of cycling while on the staff of the DOH before moving to the DPW in 2009. This official distilled evidence of the benefits of cycling from various literatures, mastered the insights and recommendations disseminated in the above-mentioned National Association of City Transportation Officials guidebooks, kept careful count of the lanes in place and in prospect in Louisville, and argued the case for giving the needs of cyclists a heavier weight within the DPW.

Changing the emphases, still more the culture, of the DPW was a struggle for reasons of both motive and opportunity. The department was dominated by engineers who (as a veteran outside observer of the DPW put it) viewed themselves as essentially mathematicians engaged in solving problems that arose in assuring the safe and efficient movement of cars and trucks on roadways. Financial sticks and carrots might of course modify behavior, and an interviewee offered the Americans with Disabilities Act as a case in point: "Under that law we've gotta design to certain standards. It has taken time, should have been done years ago, but if we don't do it we lose federal money, and that has moved things along." But no such federal imperatives attached to bike lanes.

Opportunity or, perhaps better put, bureaucratic capacity, was also less than ideal. "Lack of staff," said an official, "hinders our ability to write our own guidelines. We just don't have enough people. Our salaries aren't competitive, so new engineers come in and then jump ship to jobs that pay better, and we lose institutional knowledge." All the same, this "department of one" (as the biker/MPH/DPWer described himself) enjoyed strong support from bike advocates and, no less important, in city hall.

Jerry Abramson, mayor of the city of Louisville from 1986 until 1999 (when term limits took effect) and then of Metro Louisville from 2003 until 2011, embraced biking and active living in general. An observer recalled that Abramson, while vacationing in Boulder, Colorado, had been impressed by the vigor of cycling there and resolved to make Louisville more bike-friendly. His strategies included launching a Mayor's Healthy Hometown Movement (which included a Built Environment Committee, later retitled the Active Living Committee); convening intermittent "summits" on biking and bike–hike events on officially designated Mayor's Miles; including pro-cycling recommendations in the city's master plan; hiring an active cyclist for city hall's communications office; allocating funds for Street Sense, a program that promoted safety for cyclists and pedestrians; and extending an open door and friendly ear to the above-mentioned cyclist/engineer/MPH in the DPW.

The political antennae of active living advocates detected an important institutional implication: the future of their mission(s) might be more secure if this expanding portfolio left the hands of the DOH (which had earlier taken it over from the Department of Housing) and came under the authority of the DPW. An advocate explained:

> Championship by Public Works means money, expertise, the "E's," and the rest—they can give teeth to active living. And involvement with Public Works gets the people involved to really *see* the need for changes in the built environment—it's a kind of cooptation. DOH can't do this. They're tied up with the H1N1 virus and all that. Also DOH is not the mayor's favorite son, but he loves DPW and its director. So, though a couple of years ago I would have thought it was crazy to focus active living in DPW, now it just makes good political sense to go with them.

Unsurprisingly, cycling activists were delighted. As one put it in 2010, "It's just about the bike advocate's dream here! You've got prominent bikers working as a team within government, and the silos have broken down." The dream, however, did not then extend much beyond efforts, mainly of an educational nature, to boost the demand for active living.

Abramson's successor, Greg Fischer, himself a dedicated cyclist, took office in 2011 and not only relied heavily on the advice of the DPW's resident biking expert but also appointed a new, bike-friendly head of the DPW to replace one who an interviewee had guardedly characterized as "not not supportive" of biking projects. In 2013 Fisher also committed $300,000 of general funds for bike lanes, the installation of which no longer would take place solely in the course of repaving projects.

The money was, an observer remarked, "budget dust" compared to what was spent on roads and highways, but it offered welcome relief from the formidable hurdles that encumbered the city's search for active-living-friendly transportation funds from the federal and state governments. These barriers had three main sources. First, in largely rural Kentucky, Louisville—with its urban mores and sizable tax base—occupied, as a source put it, "another world." Cultural, economic, and political discordances made it hard to pass "policies that work for Louisville *and* the rural parts of the state. All these small counties just want aerobics classes and aren't interested in the built environment."

Second, budget hurdles grew higher after the economic downturn of 2008. The story for programs to support physical activity, a health official remarked, as for those aimed at smoking cessation (which lay much closer to the institutional heart of DOH) was the same: "sharp cuts in state funding, and less and less each year."

Imbalances of power among interest groups were a third obstacle. The clout of the road builders' lobbies, declared a cycling advocate, meant that, notwithstanding the patient effort of progressives in the state's DOT, the department continued to be "all about cars." The appropriation of $300,000 in local funds constituted a long-sought liberation from higher levels in the federal system whose responses to requests to support bike-friendly changes in the built environment usually ran the gamut from hostile to indifferent.

The confluence within Louisville of political forces favorable to biking yielded impressive results—for example, the addition of forty miles to the city's network of bikeways in 2014. All the same, weighty obstacles remained, some cultural, some strategic. In poorer communities, a source explained, "cars are a status symbol. 'Don't tell me to ride a bike!' And road reconfiguration triggers resistance among the more affluent too—such as people who commute downtown from the East Side. They hire lawyers and bog it down."

To the complexities of culture and commuting must be added the convolutions of communicating with communities. A public official recalled that

> one day we woke up to find that the Southwestern Parkway had been reconfigured and a bike lane had been added. The residents hadn't been consulted and they were livid. So the planners learned: you gotta notify and involve the community. So in the Livability Plan for the Northwestern Parkway they wanted CDC money to go from one way to two way and we held it up. We said *listen to us.* And we said, let's develop a plan together. And that plan *does* include bike lanes. And sure enough motorists *do* get mad and complain about bikers hogging the road. So we have bike discussions—be a good biker and all that. And sometimes a street gets closed for a walk/bike event—and the residents are furious.

Although frustrating and tedious, these intermittent dialogues and endless dialectics had by 2016 won Louisville a ranking of thirty-first among the nation's fifty most bikable cities, as determined from a review of diverse statistics by *Bicycling* magazine. The rankers were dismayed that the mayor's budget for biking infrastructure had been cut, that the city had been slow to implement protected bike lanes, and that a bike-share program (scheduled to launch in 2017) had been slow in coming. On the other hand, Louisville (according to a local advocate) had "gotten in the habit of reducing lane width and painting buffered bike lanes," thus enabling a growing number of wide bike lanes in the downtown, which was in turn becoming "an increasingly vibrant urban core."[12]

WALKING

Pedestrians in Louisville were less organized and assertive than cyclists, but the acknowledged importance of "walkability" in an appealing twenty-first-century downtown gave them political standing despite the diffuseness of their "interest." Nor were walkers without institutional allies. For example, at the transportation authority, an interviewee remarked, "pedestrian access is important because they want to be sure you're able to walk to their bus stops." As is customary in active living policies, however, demand-side interventions entailed less political cost and less investment of political capital than did supply-side measures. In the 2000s, therefore, the city accumulated programs and projects—for example, routes designated as Mayor's Miles, pedestrian summits, pedestrian master plans, and events and seminars addressing safety issues—that sought to encourage residents to get out and about. "Walkability assessments" that measured and rated the fitness of streets for pedestrian purposes and published the rankings on the internet and elsewhere became a prominent tool both city-wide and in neighborhoods.

Although these seemingly self-implementing initiatives generated considerable publicity, some active living advocates viewed them as more talk than action. As one put it in 2015,

> Where we've foundered is in being proactive in getting people to walk and in making infrastructure improvements. Emphasizing safety is fine, but we could do much more. The Mayor's Miles have not been all that effective. But there are some old agenda items that could come to life. For instance, post the distances to the bus stops. People will ride three blocks and then transfer! Can't we change that?! Get people to use the stairs and burn calories. Make golf courses accessible after hours to walkers. But the question is *who* should do these things? DOH? The mayor? Likewise, we've done walkability assessments, but there has been no follow-up.

Contemplating the disconnect between *envisaging* demand and *inducing* it leads beyond social messaging to the supply side of the picture—to the built environment that attracts or discourages pedestrians. Walking takes no special skill, but that elemental activity may be deterred by anxieties about menacing auto traffic and by the absence of sidewalks (or ones in good repair)—problems of design that can rarely be detached from neighborhood and municipal politics.

For example, traffic calming—regulations and redesign projects that slow the rate of speed at which cars can legally or feasibly move—is widely accepted as an encouragement to walking. Rules—for instance, lowering speed limits to twenty-five miles per hour—shift the demand side of the active living equation, whereas the most familiar supply-side strategy calls for changing one-way thoroughfares to a pair of two-directional lanes (perhaps in combination with the addition of bike lanes, buffers, and other accouterments that aim to make traffic less aggressive and more pedestrian-friendly).

None of these interventions falls painlessly into place. Lower speed limits tend to evoke grumbling by discombobulated drivers who then either comply or take their chances on a ticket. And conversion of traffic patterns along heavily used streets often evokes organized opposition not only from regular drivers on the streets in question but also by people who live on and around them. An active living advocate recalled how, in her well-off neighborhood, "it took ten years to get traffic calming—but we finally got it. And, yes, measurement showed that it did slow commutes—by an average of 14 seconds!"

Sidewalks present a distinct set of challenges. It is hard to engineer walking back into everyday life in neighborhoods that lack a network of smooth, well-connected sidewalks, especially ones that run to and from shopping areas and other heavily used venues. But getting the right kinds of sidewalks in the right places often entails heavy political combat.

For one thing, installing new sidewalks competes with fixing old ones—and with filling potholes, clearing snow, repairing damaged roadways, and other municipal fundamentals that claim the attention of the DPW. Second, the demand for sidewalks, new or improved, tends to exceed the resources the DPW budget allocates for these purposes. Therefore, walkability assessments may begin by fueling demand and end by feeding frustration. As a city councilor explained,

> [Our district] did [one]. We looked at sidewalk connectivity, surface conditions, attractiveness, and other things. We rank sidewalks on a scale from 5 to 1 and we try to fix the 5's first. We look for money for where sidewalks are missing and we work with DPW on that. We get lots of complaints and it can take years to get it done. DPW answers to 26 council districts, and they are

> all concerned with sidewalks and paving; it's a high priority. So we in the district put in the request—and there's delay after delay, and it's a source of frustration to those of us on the Council, and if the state is involved, 5 years can turn into 20.

Moreover, crafting a strategy to secure new sidewalks entails assiduous combing through local ordinances, the regulations that govern the DPW and other agencies such as Planning and Design, exemptions, waivers, and interpretations in hopes of discovering who (city, developers, residential property owners, commercial establishments), if anyone, is legally obliged to install and maintain sidewalks. An activist sketched the maze:

> The Department of Planning and Design considers applications for zoning changes, variations, and such. The Department of Public Works has to sign off on these plans—for instance, curb-cuts, traffic lights, whether there is gonna be a sidewalk, or whether they'll waive sidewalks. Planning and Design sends notices of proposed zoning changes, and I've asked them to send notices that relate to sidewalks, which they don't do now. So I've gone to many meetings where there are efforts to revise the zoning and land development codes, and I got burned out. From the developers you get "We can't do this, it's too expensive." Their mindset is the less regulation the better. But it's hard even to know what the regulations are because they're scattered all throughout the code.

The epic quest for a new sidewalk on a main street in Louisville's Clifton district illustrates the complexities and contingencies of what may be charitably termed the system. (What follows comes from two presentations by Cassandra Culin and an interview with her.) Pedestrian access had long been a concern in Clifton because the neighborhood houses the Kentucky School for the

Blind—the largest population of the blind in the United States. These residents, along with sighted shoppers, put themselves in peril when making their way to or from a supermarket, a pharmacy, and other stores along a busy four-lane road, a portion of one side of which lacked a sidewalk. In 1998 Culin, a Clifton resident who led the neighborhood in negotiations with the DPW over access issues in the community, persuaded the pharmacy to build a walkway to the sidewalk in front of its store and then unsuccessfully urged the DPW to install a crosswalk near a traffic light to make traversing the area simpler and safer.

Warming to the cause, Culin developed an organizational and political agenda in 1999. Under her prodding the Clifton Community Council added a pedestrian/bike planning committee (on which she served as coordinator) to its long-range planning process, which made recommendations for the transportation sections of the plan and designated the road in question as the most dangerous in its jurisdiction (and, hence, the committee's top priority), brought on board the district's representative on the Louisville city council, built support for the proposed sidewalk among members of the community council of Clifton Heights, and pressed the DPW to install the new sidewalk, which entailed either increasing the right-of-way or finding room within it. The advocates' initial suggestion, that the road be narrowed to make room for the sidewalk, Culin recalled, found officials at the DPW "bouncing off the walls" and insisting that the project was infeasible, but after years of meetings and weighing of alternative strategies, in 2007 the department itself proposed a "road diet" that would make room for the sidewalk by slimming the road's four lanes down to three (two for traveling and one for turns).

No sooner had the project gained official approval than the Great Recession of 2008 complicated plans for funding it. "The economy was down, gas prices were up, and capital projects were being slashed," noted a participant. The sidewalk was slated to be funded by federal stimulus funds, but it got bumped down the list of priorities by a road project and then encumbered by long negotiations with the state's Department of Transportation, which channeled the federal funds in question. Money finally in hand, the city announced that the sidewalk would proceed, whereupon bureaucratic and budgetary politics yielded center stage to the old-fashioned interest group variety. A "well-funded, vociferous opposition came out of the woodwork, with a PR person, a lawyer, and a website, and coverage spread across the most popular TV and radio stations." The volunteer supporters fought back with a website of their own—"safe42.org" to counter the opponents' "save42.org"; added allies in Crescent Hill, a neighborhood adjacent to Clifton and Clifton Heights; and mobilized a coalition that included "people with web proficiency, a certified traffic planner, a leader of the blind community, and several passionate folks to educate and rally support." Having called for a public comment period, the mayor then "stuck his neck out" and announced a final "final" approval to the sidewalk, which was installed in 2012—thirteen years after Culin had launched her campaign for it. What to do for an encore? "Near here there's an intersection that the city really botched some years ago. But they don't seem to think there have been enough injuries to justify a four-way stop! Seniors there don't feel safe crossing the road. We're working on it, but it's expensive to fix."

PARKS AND RECREATION

Walkers and bikers get physical exercise in the purposeful act of moving from here to there, but in parks and other recreational venues, the built environment also beckons to residents who might practice active living simply or mainly as part of a plan to enjoy themselves. In this respect, too, Louisville is impressively endowed. A prime example is the Louisville Loop, which, under the authority of the Metro Department of Parks and Recreation, aims eventually to offer one hundred miles (fifty of which have been completed so far) around the city for walking, strolling, running, biking, and picnicking.

The city has also sought to attract more residents and tourists to its waterfront and its waterfront park by converting to a car-free zone a bridge spanning the Ohio River into Indiana. Developed with strong support from the mayor and the city's Waterfront Development Authority, the redesigned bridge is said to be "unbelievably popular," drawing "wall to wall" crowds on weekends.

Most neighborhoods, moreover, enjoy access to municipal parks of various sizes and descriptions. District 5, a largely African American community, for instance, offers its 28,000 residents the chance to walk, bike, or engage in sports in seventeen different parks, four of them designed by Frederick Law Olmsted, the renowned creator of Central Park in New York City. A public official contended in 2015 that residents were increasingly taking up these options, citing, for example, "a walking group that was formed here. They get together to walk in the parks and even sponsor 5K walks. Awareness has been raised. People want to be healthy and these walks also let them do things with others."

Informants cautioned, however, that the use of these recreational resources was discouraged in some measure (exactly how large a measure is of course impossible to know) by the same unhappy reality that had bedeviled the ALbD planners—anxieties about safety and crime. Drugs, guns, and youth gangs combined intermittently but regularly to make headlines and to dominate the civic conversation. In spring 2015, for instance, the killing of an eighteen-month-old child by a stray bullet was widely discussed and deplored; the mayor, having attended the child's funeral, created within city hall a new office for safe and healthy neighborhoods, intended to elevate crime fighting on the municipal agenda and to bring fresh perspectives to bear on it. This new institutional attention to violence was arguably a precondition to higher levels of active living, but the one priority also traded off resources and attention with the other. "There's lots of money for violence prevention," explained an official in the DOH, and the DOH director affirmed that violence was indeed a "public health issue." "It's a bigger focus now. It's less about active living than about keeping kids in school, mentoring, and so on."

In similar vein, a high-ranking city official, asked in 2015 about the connection between active living and fear of crime, recounted how the cousin of a friend had recently dodged a stray bullet in a park only because she had happened to bend down at a crucial moment to tie her shoe. Such incidents (which cannot be "rationalized" as drug deals gone bad or violence between members of gangs) have a resonance that mere statistical trends about the trajectory of crime rates cannot match.

The official who described the near-miss in the park proceeded to note that the city had seen more shootings in the first four months of 2015 than had occurred in all of 2014. Asked about the connection between violence and the demand for active living, an interviewee illustrated how central is prevention of the former to elevation of the latter:

> A couple years ago middle school kids in gangs had a fight and one kid died. A vigil was held on the waterfront and a rival gang

> showed up. There was a fight that spread downtown into the arts district and did some damage. The waterfront is a big deal for us. There have been incidents on the Big Four [bridge over the Ohio River] too. These things create a public perception that the waterfront and the parks are not safe. That's bad because the parks can be the only place people have to exercise. These people aren't buying gym memberships!

In Wilkes-Barre (discussed in the previous chapter), support for active living was concentrated in two distinct institutional venues—business groups working to redevelop the downtown and trail associations seeking to extend and connect trails around the city—and focused mainly on walking. Louisville, by contrast, found formidable advocates for active living in a range of public, private, and voluntary organizations variously engaged in the promotion of walking, biking, and recreation. In Louisville, commitment to active living was, so to speak, a mile wide but an inch deep. In 2007 a veteran of active living initiatives noted that "we have a few allies in every department, but you gotta find the right ones. Old-fashioned bureaucracies are all around. They've always done it thus and so even if it makes no sense and changing that is a heavy lift."

In 2000–2015 advocates of active living worked to achieve greater depth, which meant transcending active living committees and networks of multiple members with varying degrees of commitment and finding a secure and hospitable institutional home that could infuse coherence and energy into the bits and pieces of policy at hand and under development. Over many diverse but disjointed accomplishments, a veteran of active living politics contended, loomed one big question: "Who should do what?" The political scene offered no entirely satisfying answers.

One might expect to find a comfortable institutional home for active living in the Department of Health. As noted above, however, in 2010 the department ceded leadership of that portfolio to the Department of Public Works. That active living, the objective of which is to promote health, fit awkwardly within the health department is explained partly by bureaucratic traditionalism. A source contended that "below the top leadership the staff is very closed minded. There are program categories, like food inspection and such, and the program people like routine. They don't think about public health systematically. So they resist innovations like active living."

The vigor with which leaders of the DOH counter the resistance to a deeper engagement with active living depends, in turn, on how they happen to view the central mission of public health. The director in office in 2008 "got" both active living and its attachment to the built environment and allocated staff time accordingly. His successor was said to hold a more traditional image of the duties of a public health agency (smoking cessation, for example) and a conviction that violence deserved more prominence as a public health problem—a combination that pushed active living and the built environment down the list of departmental priorities. "We've got no staff to devote to it here," a departmental staffer remarked in 2015, and so the engagement of the DOH with active living projects had grown "remote and tangential."

The Department of Public Works saw a reverse evolution: an enthusiast for active living replaced one who had been less so as head of the agency, which allowed the department's able and energetic active living aficionado greater latitude to craft and pursue initiatives. But the department, which remained largely a preserve of tradition-minded engineers, needed (as various observers remarked) a culture change that would include the hiring of new staff trained to address the engineering challenges that accompany attempts to make biking and walking safe and appealing. As noted above, the stirrings of cultural change were discernible both in the dissemination of authoritative alternatives to the manuals and guidelines traditionally honored by highway and traffic engineers and in a diffuse sense within the profession that the rules of the game for the built environment were changing. But how far these tendencies would inform agency behavior depended heavily on encouragement and pressure from political champions, especially in city hall.

Both Mayor Jerry Abramson and his successor, Greg Fischer, strongly endorsed active living, and for understandable reasons. As a biking activist explained, "Active living initiatives are important in an economic development strategy. There's a competition among cities to be more bike friendly. Young executive talent doesn't want a one-hour commute and may not even want to own a car. You bike to work, you stay fit and healthy." A public official concurred: "The mayor wants us to be a top 'healthiest city.' Unhealthy cities don't attract businesses." But city hall had many other demands to meet, some of them routine and predictable (protection against fires, cleaning streets, removing snow, and more), some unforeseen and urgent (responding vigorously to the above-noted killing of an eighteen-month-old by a stray bullet, for example), and these compete for mayoral attention and resources. Declarations and events that aim to galvanize demand for active living—the Mayor's Healthy Hometown Movement, Mayor's Mile designations, pedestrian and cycling summits, and the like—understandably take precedence over supply-side initiatives to change the built environment, which tend to be expensive, dependent on multiple bureaucratic clearances, and opposed by segments of communities whose familiar spatial arrangements may be disturbed.

Money, to be sure, is a potent motivator, and the prospect of a federal grant to modify the built environment reliably attracts favorable attention in city hall and the agencies. But such awards are often mediated by state authorities, which, several sources opined, not only worry mostly about bridges, cars, and roads and regard Louisville as an alien metropolitan presence in a largely rural state but also move at a glacial pace. Grants that came directly to the city—from the CDC or the Robert Wood Johnson Foundation, for example—were highly prized but too ad hoc and project specific to unite the institutional fragments that govern Louisville's built environment.

Representatives elected to the twenty-six-member city council vary widely in their commitment to active living. Some confine themselves to contextual supports such as badgering the DPW to repair cracked sidewalks, while others promote walking clubs and walking and biking events in their districts. Some view changes in the built environment, such as new bike lanes, as an improvement; others fear them as a trigger for conflicts into which they will be drawn and which they will be expected somehow to manage. Nor do enhanced opportunities for active living stand high among residents' priorities. In one mixed-income community, a representative explained, the top concerns are "crime, safety, and cleanliness; jobs; and respect for the community—not littering, not throwing stuff out of the windows, and so on."

Louisville residents do not lack formal opportunities to shape the built environment around them—for example, by seeking to influence the transportation sections of the neighborhood plans, perpetually in development, that find their way into master plans such as Louisville 2020—but participation tends to be lackluster. The inertia that dogs the practice of active living also discourages deliberation on it at the grass roots. As an elected official recalled, "We were supposed to form an active living committee in our district, but it hasn't happened. Active living is not an easy sell. If you frame it wrong, eyes glaze over, people are too busy, and so on."

The possibility that active living might lower health care spending is not lost on local companies and insurers, and so private organizations increasingly incorporate physical exercise into wellness programs for their employees and will sometimes donate money to active living events. Humana, for instance, created a bike-share program for its workers, a local foundation sponsored a "ride to beat cancer," and employers were said to be warming to discounted insurance premiums for workers who worked out at the YMCA. But these heavyweight organizations seldom join the fray over proposed changes in the built environment and remain content with their status, as a health official said in 2010, as "detached . . . minor players." Detachment, moreover, seems to grow in inverse proportion to the income levels of the city's neighborhoods, a state of affairs the geographical irony of which was not lost on an interviewee: "The health institutions are very often *in* these communities, but they don't know what to do there. Their hearts aren't in it. The hospital people and other staffs are scared of these neighborhoods." Meanwhile, the fitness and sports programs run by voluntary organizations such as the YMCA and the Urban League are constrained by limited budgets that must stretch across multiple projects.

Finally, the ad hoc and incremental nature of advances in active living in Louisville reflects not only the diffuse political distribution of its proponents but also, and perhaps mainly, the ingrained bureaucratic culture of the Federal Highway Administration and the entrenched power of cohesive stakeholders—developers, highway builders, real estate agents, and other beneficiaries of the car culture—who stand ready to summon legal arguments, economic computations, and political clout to keep inimical proposals off the civic agenda or to squelch them before they get far. Recapitulating more than a decade of activism, a transportation expert in Louisville remarked ruefully:

> In the big picture what we've done is small potatoes. The feds, the planners, the developers are interested in keeping the car culture. They want to build out into the green fields, they want the highways, they want the bridge projects that just do harm, just suck people out to the suburbs. And they don't care about pedestrians and cyclists. *These* are the big stumbling blocks.

To be sure, countervailing economic considerations—for instance, the lure of walkable and bikable communities to the young, well-educated workers that mayors want in their jurisdictions, and the cresting conventional wisdom that opportunities to walk and bike boost home values in new (and renovated) housing developments—combine with the growth in numbers and engagement of active-living-friendly organizations to enhance countervailing political power.

Such power, however, may rise today and fall tomorrow. Active living is perpetually "on" most civic and organizational agendas in Louisville, but its standing on those agendas is forever vulnerable to what one interviewee called the "issue du jour problem."

> With the ALbD grant, housing was the issue du jour, and we tied it to active living. Then along came another big Hope VI grant

> for a housing project and the hook this time was environmental greenness—it was gonna be smoke-free, there'd be composting, and such. But active living was not part of it! Then there was another big project, and it was all about community interaction—and not active living! DOH has lately done a project that aims to advance health and social justice, and equity, and yes, that's a health focus—but not active living!

In its struggle to ascend and persist as a high political and organizational priority, active living simultaneously enjoys the blessing of consensus and suffers the curse of complexity. Supply-side designs to recast the built environment routinely yield to demand-side exhortations to get exercise when, where, and as one can, and these amorphous urgings in turn give way to other issues du jour, such as healthy eating (which can advance by means of policy levers such as taxation and regulation, mostly unavailable to active living). As these policy tides ebb and flow, it remains unclear how Louisville might forge its impressive policy parts into a cohesive whole designed to boost levels of physical exercise among its residents.

3

ALBUQUERQUE

Reshaping a Cultural Landscape

Albuquerque, New Mexico's largest city (558,000 residents), anchors a large metropolitan region (population 903,000) sprawling out from a downtown that was, in the first half of the twentieth century, a proud point of passage along fabled Route 66. Close by the Sandia Mountains, the city cherishes a civic culture replete with Old Western values such as wide-open spaces and a resistance to public planning, and features a built environment shaped by the interplay between the dynamics of real estate markets and the public infrastructure (roads, bridges, and sewer lines, for example) required to make those private markets profitable. A city in which mass transit entails little more than sporadic bus service (except along the Central Avenue corridor) and in which cars are generally indispensable for both work and play, Albuquerque would seem to be an unpromising venue for the promotion of active living by means of public policies, especially ones that require reconfiguration of the built environment.

This simplified summary of the prospects for active living policies in Albuquerque, although accurate enough at the start of the twenty-first century, became within a decade and half severely simplistic. This chapter explores the sources and consequences of change in the city's imposing physical—and cultural—landscapes.

THE NEW MEXICO MODEL

At the turn of the twenty-first century opportunities for active living, including a conducive built environment, in Albuquerque were shaped by an imbalanced political economy in which the public sector contentedly played cheerleader to bankers, lenders, real estate agents, contractors, and developers. "The development community, and especially the sprawl development community," an active living advocate explained, "is a very, very, powerful special interest, the eight-hundred-pound gorilla of land use policy." This power derives importantly from "the New Mexico model, which equates rooftops and economic development, so it's sacrilege to say we should slow the spread of sprawl or have a moratorium on development." Critics of the model also cited the baleful logic of campaign finance: "The mayor [who held office from 1993 to 1997 and again between 2001 and 2009] is totally bankrolled by developers and has a favorite saying—'the only thing people hate more than sprawl is density.' He really loves that line." The prevailing mindset, another proponent of active living added, is "'we're not Portland—or Manhattan. Density will be bad for us.' You say 'density' and the immediate reaction is: 'What about my view?'"

This cultural landscape encouraged a presumption of laissez-faire toward the physical landscape that left leaders of the Alliance, the organization funded in 2002 by the Active Living by Design (ALbD) program of the Robert Wood Johnson Foundation, initially at a loss for points of entry and leverage. Said an ALbD activist:

> We're "just leave us alone" types, with our single-family homes and cars and the rest. Just let us do what we do. Build more subdivisions and more people like us will move here. And this approach has shifted a low-income agricultural lifestyle to the suburban edges of the city in cookie-cutter developments with terrible crime. "Land rich, cash poor" describes a lot of the valley, and with water very desirable, developers are buying up the land. So it's "let the chips fall where they may" and above all resist government planning.

Even some progressives, who favored smart growth, acknowledged the force of the case for little-fettered development. An aide to a county commissioner explained: "We *do* want employers to come in. Our district is the oldest, poorest part of the city, with high unemployment, high drop-out rates, high rates of teen pregnancies, and more. So good jobs and training mean a lot. We favor smart growth, but there's the rub."

Proponents of smart growth were often said to be too "pragmatic" and daunted by the deep cultural roots of the New Mexico model to stand firmly against the real estate agents, developers, and builders who profit from sprawl. As a self-described progressive explained, "It's the democracy issue. It's the West and the cowboy mentality prevails. It's the land of opportunity. People making $30,000–$40,000 a year want that two-thousand-square-foot house with the yard and the two-car garage. So when you get asked 'Why are you against growth?' it's not easy to answer."

Answers were, in fact, close at hand: those sprawling homes in cookie-cutter developments depend heavily on tax-based subsidies for infrastructure and public services, and the model obliges homeowners to use cars to get to work, play, and shopping. Still, the cultural force of the model—encapsulated by one source as "We like open space and we don't want to be on top of our neighbors"—looked highly durable, as did the citizenry's attachment to the automobile. As one exasperated critic put it in 2008, "In New Mexico, obesity is a huge issue, with diabetes and all that goes with it. But people won't get out of their cars. The only people who walk in Albuquerque are transients, poor immigrants, some local neighborhood folks, and DUI's who had their cars impounded."

PLANNERS

Albuquerque's city planning department housed a small core of critics of the reigning model who, unsurprisingly, portrayed their efforts to introduce change as a sharply uphill climb. Lamenting that "it's hard to get even the most minimal stuff to move forward," one city planner sketched the multipronged obstacles: "inertia, multiple rules on the books, and arcane manuals and city specs," which encouraged

> walling in subdivisions like feudal villages—but with no stores. The way we design streets in Albuquerque is a chicken-and-egg problem: the developments have walls so they don't face the horrible street. So why change the street? We're going backwards. The mayor and others at high levels either haven't got the information or don't see it as sexy enough to change it. They say the market tells developers what to do. But if people want a house, they are open to options. But they end up with three cars, no buses, and no chance to walk. The developers do the subdivisions, and they have the builders and Realtors working with them. The consumers are uneducated. Once they're in the house they ask

> "where's the park?" but by then it's too late. It's like a Maslow hierarchy: shelter, food, clothing, and then community.

Downtown redevelopment had fared little better than its sprawling suburban complement. "We wanted to attract younger people downtown, so we put up bars and movie theaters and housing that is too expensive. So what we attracted was a lot of young drunks."

The resistance of traffic and highway engineers within the city, county, and state governments reinforced private power. For one thing, a planner contended, planners and engineers have distinct professional mindsets. "Planners are more comprehensively educated. They study lots of stuff and take a holistic point of view. The education of engineers is very narrow, and their view is 'we're moving cars, who cares about sidewalks?'" "They don't want to do things differently," echoed another planner. "Some are just unwilling to change, and some don't want to redo the blueprints on their computers. They view themselves as a profession of their own."

Moreover, engineers were said to use risks of liability as an all-purpose excuse for inaction. "If we approve X, it's gotta be safe according to the standards of the city, the state, the feds, and so on," a planner complained, and a colleague then offered a case in point: "In this department we've been advocating for pedestrians for decades. I recently suggested we plant trees between curbs and sidewalks. The engineers replied that cars would hit them and the city would get sued. I said let the city sue the driver for damaging public property! You have to persist. You gotta push to implement these things."

Yet another barrier to the planners' innovations was the constrained size of the city workforce. Although no fan of the engineers, a planner conceded that "they are overworked: one woman looks at all the plans for all the projects for designs from all the developers. And the developers push the city to be quick, so there's no time to breathe! They don't have time to keep up with new ideas. The system is rotten!"

All the same, the planners discerned among the citizenry some emergent taste for change in their built environment. Summarizing a public meeting attended by 250 people to discuss what people liked and disliked about their city, a planner highlighted complaints about ugly highways, congested streets, and ubiquitous commercial strips, adding a widespread wish that Albuquerque could be made to look more like Santa Fe. "Form-based codes," planners suggested, would move the city in that direction, and "the city council is looking at that. But change suffers from a lack of trust of government in general, so they invited the private sector in, and in no time they came out of the woodwork and opposed it. They prefer the status quo and fear the unknown." As a city councilor remarked: "It's hard to deal with the industry on technicalities of a form-based code when they're so suspicious. They think government will put a knife in their back if they give it the chance."

Planners and their allies continued to argue that the city's decades-long incremental accumulation of zoning rules, subdivision ordinances, and ad hoc exceptions should all be harmonized under the authority of a comprehensive general plan. But each step in that direction was checked by a dispiriting interplay of private intransigence and public diffidence. For example, a source explained that

> the Environmental Planning Commission—that's the crux of the matter for pedestrians. The subdivision ordinance needs to be completely revamped. Likewise, the standards in the

> Development Process Manual for streets, drainage, hydrology, and so on. But if the commission moved to do so, the developers would be in an uproar and so the commission won't bite the bullet. Meanwhile, the Development Review Board can waive anything. So I said why not put an urban designer on the board, but there was resistance from the engineers, who didn't want to rock the boat. That board is insulated, and no one is gonna challenge the engineers.

SCHOOLS

In another corner of the institutional environment, proponents of active living tried to persuade Albuquerque's public school system to play a larger role in reengineering walking and biking back into the daily lives of their young charges. The most visible way to do so—federally funded at that—was to create Safe Routes to School (SRS). As was the case with virtually all active living initiatives, no one in the schools (or elsewhere) thought this a bad idea, but putting it in motion was a heavy lift politically. The Albuquerque school system is almost entirely independent of the city and county governments: voters elect a school board, whose members appoint a superintendent, who is responsible for the selection of principals, who take cues not only from their formal superiors but also from parents, teachers, and from state rules governing transportation, physical education, and other topics.

In 2008 active living advocates identified several barriers to implementing SRS and kindred initiatives such as the walking school bus (a group of children walking to school with one or more adults). A large barrier, ironically, was creating a favorable built environment in and of the schools themselves. A school official explained that

> we do school siting right here in this office, and we'd like to promote SRS. But the reality is we locate new schools where the land is cheapest. And there is a disconnect between the city master plans and walkable communities. We *do* work with the city planners and they approve sector plans for parts of the city. They're nice maps—a school here, here, here. But in a market driven by private development, who's gonna give us the land? So we buy where we can.

An active living advocate on the city council concurred, but in sharper terms: the schools' "insular process for new school location has indeed led to more sprawl." Walking and biking "are not a consideration at all. In fact, even good automobile planning is ignored by Albuquerque Public Schools until it results in problems that are the left up to the city to resolve."

A second constraint is the above-noted state regulations on transportation to school. "We have little leeway," a school official explained, "and we follow state rules to a tee." Third, innovations encompassing schools, kids, and risk of injury trigger concerns about liability. "One school wanted to do a walking school bus," an official remarked, "but the liability is a huge issue. We've got crazy drivers down here, and it's scary."

Even if siting, state regulations, and liability could be set aside, however, individual schools and the network of principals, parents, and teachers who staff and steer them have the first and last words as to whether SRS and other such innovations will proceed. An official in the central administration of the school system remarked that although she and her colleagues who worked on wellness were "passionate" about SRS, they could not " implement the program per se in the schools"; the official then illustrated the challenges by referencing the positive reception a national expert enjoyed when he encouraged SRS and other active living measures during his visit to an elementary school in the city:

> He generated a lot of energy and discussion. The parents seemed to want it. Then there was a principals' shuffle; that one moved on and we had to start over. For SRS you really need local advocacy. The parents and the principal have to be all for it, have to take the lead, and have to follow through. We here in central administration deal with all the schools, and what you need is the right person right there to push it. A couple schools do have strong principals, and some parents, to support it, but turnover is such a problem, and relationships are the key.

SUGGESTIONS FOR—AND OF—CHANGE

Although "uphill struggle" was the dominant motif for proponents of active living in Albuquerque in 2008, virtually everyone interviewed also remarked hopefully on signs of progress. First among these signs were various new enactments and projects scattered across the policy terrain. The Parks Department was working with leaders of the ALbD program to create "walking loops" and "prescription trails"—designated paths within parks on which physicians could encourage their patients to walk.[1] The city had adopted "impact fees" that were reduced for development projects that, instead of (or along with) building at the edges of Albuquerque, aimed to revitalize parts of the inner city. A city planner noted that "a long and frustrating" process had yielded "a new subdivision ordinance, a street ordinance, a sidewalk ordinance, and a development process manual." The city council had slowly and in the face of great resistance, drafted a version of form-based codes. The city and county were working with the federal conservancy district to convert dangerous and unsightly irrigation ditches into walking paths. And, not least important, the contentious concept of Complete Streets had advanced to the stage of public hearings.

Although disjointed, disputed, and dispersed across the policy scene, these initiatives bore witness to incremental modifications in the civic mindset that framed the perceived benefits and costs of active living. The head of an advocacy group that focused on "our physical environment and how to improve it" observed that the smart growth movement had "got more interested in public health, in how a poor built environment affects people's health, creating synergy for new partnerships and a fresh ideology that is not just 'do better growth and planning,' but rather 'this is killing us, it's unhealthy for our kids, it causes obesity and diabetes and so on.'" Reciprocally, an ALbD leader explained, public health showed growing interest in smart growth and climate change: "We hope more walking will generate a reduction in BMTs [British Thermal Units] at the regional level. Changes in land use and transportation policies are where you get the biggest bang for the buck in lowering them. There's less money now in the highway trust fund, so the answer now can't be more asphalt. We're seeing a great convergence of issues."

A veteran city planner conceded that since the 1990s "the development community had gained power partly because they contribute heavily to election campaigns" but then opined that "there's been some change in the last five years because the public has become more aware of the environment," an awareness resulting from greater attention to global warming, climate change, and rising gas prices. Even the developer-friendly mayor was beginning to talk the talk. "He never used to listen to the planners," remarked one of that stoical tribe, "but [an ALbD leader] got others to talk to him, and whoever visited him rang a bell." Distrustful progressives who listened in amazement to the mayor's newly minted references to a "green city" and "Great Streets" complained that he was "bastardizing" New Urbanist lingo.

These emerging shifts in mindset both reflected and reinforced the rise and maturation of new organizational structures and networks that worked to secure a place for active living in broader strategic portfolios. Depicting the gradual growth of countervailing power, a progressive noted that "a growing coalition of folks," such as 1000 Friends of New Mexico (which was "effective in framing a vision of how public money should help to shape development") and New Mexico Voices for Children, was bolstering the capacity of forces opposed to market-driven policies of sprawl and laissez-faire.

On this score the ALbD grant was, to mangle a poetic conceit of Wallace Stevens, a most appropriate addition at a most propitious time. As a planner explained, the director of the Alliance (the organization formed by the grant)

> has got the health community and the planning community together for the first time. Usually one side doesn't know who the other is. For instance, at the University of New Mexico you've got people who have worked on pedestrian and walking issues for years and don't know another similar group at the same university! The Alliance is connecting organizations, not just people, who come and go. It's creating institutional connectivity, which is hard to do.

The Alliance enabled critics of the New Mexico model among county commissioners and city councilors and their staffs, city planners, staff on the Mid-Region Council of Governments (MRCOG), Walk Albuquerque, 1000 Friends of New Mexico, the University of New Mexico, and other organizations to coalesce around strategies for change that both advanced a range of objectives and pushed active living higher on diverse organizational agendas.

As these organizational networks grew broader and deeper, they began to incubate and implement new political strategies. Advocates of planned development sought out allies in neighborhood associations whose members, a legislative staffer explained, "feel threatened by New Urbanism but know there's development pressure and want it to be quality development." The mayor, as noted above, now at least listened to the case for "quality." Planners were enlisting support among a more heterogeneous cast of organizational characters and refining their tools of political persuasion. One cited the redesign of an interstate corridor as a good example of how to do it:

> We lined up the Chamber of Commerce and the Economic Forum in favor of it and got the Homebuilders Association to support it too. At city council hearings, the dean of the university's school of architecture spoke for the project and I met with many council members beforehand. In one-on-one meetings like that they [council members] can ask questions, and there's a higher comfort level.

Perhaps the most vivid evidence of the growing comfort of Albuquerque policymakers with open deliberation about the built environment was elevation to the formal public agenda of the proposal to explore the creation of Complete (initially dubbed "Great") Streets, the meaning and merits of which were explored in a public strategy session held on April 7, 2008, in preparation for hearings before the Environmental Planning Commission three weeks later—hearings that would, proponents hoped, generate positive recommendations to the city council.

Streets so denominated, an advocate explained, are "urbane," create "a sense of place," and are "safe, visually attractive, and economically vibrant." In a setting so endowed, "people don't think twice about walking down the street to reach a destination or a transit stop."[2] They would give Albuquerque, which was at once "a big city and small town," the "urban experience" its citizens admired in Santa Fe (or in the city's own Nob Hill neighborhood, for that matter). A two-page draft of the Great Streets Facility Plan, authored by the city's Planning Department and made available before and during the meeting, sketched the essentials. Great Streets are "memorable" ones that encourage mixed-use development and create a more balanced and equitable allocation of space among pedestrian, roadway, and private (residential and commercial) "realms," and for the walkers, cyclists, motorists, and transit riders who inhabit those realms. The draft enumerated five "community benefits" of the strategy—enhancing "sense of place," improving access for "all travel modes," providing better safety and security, economic development, and, in succinct homage to the rising prominence of active living in the policy gestalt, "[contributing] to a healthy community."

CHALLENGING THE NEW MEXICO MODEL

As Albuquerque entered the second decade of the twenty-first century, active living advocates were ambivalent as to whether the city was progressing incrementally or marching in place.

When this question was posed to interviewees in 2010, it repeatedly evoked a "yes, but . . ." response. A city planner observed that the character of local conversation about the built environment was starting to change: "The best thing lately is that the vocabulary of the [Complete Streets] plan is now more and more widely used. It's become part of how we discuss city planning." Moreover, advocates had continued to refine political strategies to get those who had begun to talk in a new way about active living and the built environment to walk the proverbial walk. For instance, an activist explained how planners and supporters of active living were making their case to city councilors by encouraging their aides to attend meetings of neighborhood associations at which the head of the ALbD project presented visuals of Complete Streets and led discussions of what that increasingly familiar locution might mean for local residents.

Progress was halting, however. Developers warned about the costs that tougher regulations would entail, and the city council remained short of a majority to enact legislation authorizing Complete Streets. Form-based building codes had been approved by the city council (which, an admiring proponent of active living remarked, had used its own staff, not the city planners, to draft them). But the codes were elective, not required, and therefore left it to developers whether to follow them or not. A strategic action plan had been published for southwest Albuquerque and, although some progressive features had survived amid many deletions, myriad antiquated zoning rules lingered on the books, and the obduracy of city engineers still hindered those seeking to make the built environment more amenable to active living. Recounting her struggles to improve for pedestrians a very "compromised" corner at a busy intersection, a planner lamented: "The engineers are a sad story. City policy says these intersections should be thus and so, but the engineers don't take it seriously. When it comes to signals, for example, they'll only look at it from the point of view of the cars going through, never mind the pedestrians, the bikes, the seniors trying to get around there. That just doesn't cut any ice with them."

Prescription trails, developed under the aegis of the state's Diabetes Prevention and Control Program, were becoming more numerous and more widely used. Guides to the park paths that lent themselves to easy walking were being published;[3] the newspapers were covering them; federal, state, and local park officials, eager to bring favorable public notice to resources that some deemed "non-essential" in tough budgetary times, were collaborating; and the promoters of trails had discovered a warm and fuzzy hook for their efforts—namely, the promise that walking the trails could help prevent or manage diabetes in dogs as well as in their owners. All the same, providers who were expected to "prescribe" and encourage use of the trails were said to be "leery" of doing so and "stressed." "They said it's hard to incorporate into their practices," explained a leader of the effort, "so now we ask whom should we target in the health care professions? A shotgun approach is no good, and so we're looking at primary care providers, nurse practitioners, certified diabetes educators, and maybe pharmacies."

Interest in SRS was increasing too. A school official recalled that when the head of the active living Alliance had inquired a couple years earlier how many schools were prepared to give it a try, the answer had been "zero," but more recently parents and principals in three schools had indicated tentative interest. Still, in a school that found SRS too ambitious and pursued a walking school bus instead, this more modest project had foundered because, to ensure safety and avert liability, the parents leading the "bus" had been required to undergo background checks. Some parents had been deterred by the fee imposed to get this clearance, others by anxiety over possible disclosure of irregularities in their immigration status. Meanwhile, active living enthusiasts, who were trying to improve the dialogue between officials in charge of the schools' capital master plan and city engineers, were heartened when one of the latter camp attended a meeting of the former—and then were disheartened when he never came to another, claiming to be too busy. Clearly, much was in the air. But whether and when (if ever) these strategic vapors might crystallize into policies that would change the built environment remained anyone's guess.

Above and beyond this incremental/decremental two-step, however, a contingent coincidence of political economy in the fall of 2008 was reshaping the policy environment in which the frustrating show transpired. On the political side, a split within Democratic ranks in the mayoral election of 2008 gave Albuquerque its first Republican mayor in decades, Richard Berry. Advocates for active living pointed skeptically to his party affiliation and his background in business but were soon pleasantly surprised by his pragmatism, the quality of his appointments to city hall and within city departments, and his willingness to meet and work with the city council—executive virtues they rarely ascribed to the previous incumbent.

Local economic ripples from the national financial collapse and recession that arrived around Election Day in 2008 tested those mayoral virtues immediately and for years to come, for the downturn undermined confidence in the merits of the New Mexico model of development. Escalating unemployment, foreclosed homes, and elevated gas prices raised the costs of sprawl and thereby called freshly into question the sustainability of a laissez-faire approach to metropolitan growth. "The financial collapse has put the brakes on sprawl," a legislative aide tersely declared. That more residents began leaving cars at home and taking the bus was hardly a ringing endorsement of density, but the economic exigencies amid which the new mayor took office intensified attention to, and the appeals of, the script that active living proponents sought to sell to public and private producers of the built environment.

THE CONSOLIDATION OF COUNTERVAILING POWER

In mid 2008 unemployment in Albuquerque ran around 45 percent, a figure the financial crash drove up to 8 percent one year later. In 2010 it remained at 8 percent and receded slowly thereafter. In June 2015 New Mexico's 7.2 percent unemployment rate was surpassed nationally only by West Virginia's, and Albuquerque's rate, at 6.4 percent, was not much better. The city faced a double assault: on one front, stagnation in the private economy meant sharp downturns in construction; declines in home values, prices, and sales; and a rise in defaults on mortgages. On another front, the public sector, a prime employer in the region, was retrenching—especially the federal government, which cut back military bases and laboratories to the tune of, by one estimate, two thousand jobs. Albuquerque's urban economy, warned one observer dramatically, was "on life support."[4]

In these straitened circumstances, calls for economic revitalization took on new urgency and attracted wider and more attentive audiences than had been the case before the collapse. One school of thought, especially prominent in the outlying unincorporated areas of Bernalillo County, counseled doubling down on the familiar laissez-faire approach: strive to attract new employers, whatever the quality of the jobs and however much regulatory indulgence it took to get them. The usual suspects in support of New Urbanism, smart growth, sustainability, and active living countered this endorsement of sprawl by pointing to a "double demographic" trend the economic downturn had highlighted. On the one hand, young people—millennials—were said to be abandoning Albuquerque for more cosmopolitan and congenial city settings such as Austin, Portland, and Denver, which (as interviewees remarked) had "vibrant activity centers"—"swingin'" downtowns with walkable and bikable places that enabled residents to work, play, and shop conveniently and without need of a car—and fostered opportunities for entrepreneurship, innovation, and small businesses such as galleries and microbreweries. Simultaneously, boomers approaching retirement age were said to find Albuquerque wanting as a site in which to age in place in affordable downtown housing with easy proximity to a range of attractive activities and amenities. The progressives, a participant noted, carefully avoided coupling their agendas with endorsements of density, which remained a dirty word in Albuquerque, and instead crafted and embellished an "urbane" image of the city that found new resonance as an antidote to the economic doldrums.

Electoral shifts reinforced economic pressures. Perhaps because Berry was the first mayor to be chosen under rules, adopted by referendum in 2005, that created an "open and ethical election fund" to help finance local elections (which made the city's new Republican mayor less beholden to developers and other big financial contributors) or perhaps for other reasons, most interviewees agreed that he "got it." As one source remarked, "He sees that we need to improve downtown, make it more walkable, more urban, because we're losing our young people. Millennials don't want to be here without a real urban environment."

Moreover, new staff in city hall reflected the mayor's openness to innovation. Savoring the irony that a Republican businessman would pull in a Democratic "cadre of the creative class," a political insider ticked off their roles—for example, a "rainmaker" with good connections to foundations; a community organizer from the youth development movement who enjoyed "street cred," especially with Latinos; and an African American poet who acted as "cultural ambassador" to and beyond the Black community—all of them invaluable to "rebranding the city as a center for innovation and entrepreneurship."

City hall was not the sole source of new political energy on behalf of active living. The election in 2012 of a progressive nine-year veteran of the city council who had successfully promoted active-living-friendly projects within her district to the board of county commissioners (of which she became chair) much elevated such initiatives on the agenda of Bernalillo County, a heavyweight member of the MRCOG of New Mexico, through which pass federal transportation funds for Albuquerque and its surrounding counties. And although only three of the city's council's nine members were said to identify closely with active living and the priorities (such as mixed use) that accompanied it, fresh winds blowing from city hall and the county reframed their work. A planner explained how one councilor had, for example, used budgetary set asides, which members could access to fund individual projects, to introduce fewer lanes of traffic, buffered bike lanes, and wider sidewalks on a major city thoroughfare, and traffic calmed by means of road diets on other nearby streets. Such concrete pictures of change in the built environment, abetted by the increased use of visuals in presentations to neighborhood groups, doubtless added value to the progressive pronouncements now more frequently emanating from city hall, the city council, the county commission, and the regional planning agency.

These departures in political economy modified the city's cultural landscape—how residents and leaders envisioned and discussed the city's identity and future prospects—and thereby shifted the balance of power between skeptics and proponents of active-living-friendly changes in the built environment. The first and most palpable result was the mobilization of the mayor, the business community, and the city council behind the long-lingering Complete Streets legislation, which the council approved by unanimous vote in January 2015. The measure requires "streets that are designed and built to efficiently serve all users, including pedestrians, cyclists, transit riders and motorists."[5] This new legal framework emboldened progressives in agencies in the city and the county (which adopted its own version of Complete Streets in June 2015) to intensify their efforts on behalf of reforms they had long been struggling to promote.

A prominent example of this bureaucratic élan featured the efforts of city planners to go beyond Complete Streets to an encompassing Integrated Development Ordinance (IDO) for the city. Planners who had for decades pushed vainly for rules supportive of multimodal development and transportation, which many had despaired of seeing in their professional lifetimes, now sought to harmonize within a comprehensive plan the hundreds of area-specific "plans" that had been adopted piecemeal over many years at the behest of neighborhood voices seeking exceptions to official zoning and other municipal regulations. (Indeed, the term "Uniform Development Ordinance," used at the outset, had been changed to "Integrated Development Ordinance" precisely to meet objections about city-enforced sameness.)

The process began in 2014 when a new director of planning, charged by the mayor with reviewing the city's zoning processes, asked a veteran within the planning agency why it was all so complicated and how to fix it. The reply, the planner recalled in an interview, was that

> we gotta start over. We've got general plans and more specific area ones, which, back in the '70s, were supposed to be few, mainly to protect the character of, say, three-hundred-year-old neighborhoods. But they became a political tool to address constituent concerns and had no overall rationale. There's now a huge mish-mash of plans, with no coordination, and few get updated. Neighborhood associations cling to these plans like life preservers. They see them as protections and get mad at us if we disagree. The system is unworkable to enforce and interpret, and there have been different interpretations and lots of amendments over time. So different stakeholders read from different versions. So that's what I said to the director: we need to get it structured and clear. A citizen shouldn't need to hire a lawyer to go before the planning board. If we do this, we can implement our vision of centers and corridors and mixed use and protect the single-family homes too.

As the distinct new modules of the IDO were unveiled in 2015, their designers expected—and faced—opposition less from developers (most of whom welcomed clarification of the rules of the game) than from neighborhood associations (described by one observer as "self-appointed protectors of the public realm"), which were often in fact little more than single-member veto groups that nonetheless sometimes succeeded in rallying progressives in defense of "the community." Reprising the retail political strategy that had helped them to enact Complete Streets

("We talked individually with each city councilman and county commissioner to get them on board"), gradually built support among the public and political leaders. In 2018 the new IDO, passed in 2017 and amended twice thereafter, took effect.[6]

Growing support for active living initiatives came from above as well as within the city as the regional planning body, the above-mentioned MRCOG, ceased merely (as one source put it) to "just predict sprawl forever" and instead began to link transportation policy to other policy arenas, including health. As early as 2010, a staffer recalled, the MRCOG reached out to "the health community" in preparing its regional plan for 2035. In 2011 the link grew stronger when an institutional member of that health community, the New Mexico Healthier Weight Council, founded a Complete Streets Leadership Team. In 2012 it grew stronger still with the arrival of a new progressive chair, who had (as noted above) pressed for active living projects when serving on the city council. Asked why the agency expanded its agenda to include active living, a staffer replied that "the environment changed. Several of our staff had a strong professional or personal interest in active transportation, so we naturally expanded our scope. And the leadership here is flexible and let us explore what would work."

A staffer recounted the agency's evolution from its "cursory look at how we might do things differently" in the plan for 2035 to more ambitious "scenario planning":

> Looking at climate change, we took up flood control, and that led us to density versus sprawl. That started a conversation, which led to preferred scenarios and recommendations that we work toward them. We presented the region's official forecast—the path we're on if we don't do things differently—and contrasted that with the preferred scenarios and the policies that sustained

> them. Back in 2010 we didn't even talk about climate change. Now there's a public health section in our long-range plan.

MRCOG, an advisory body whose "forte," a staffer explained, is "offering perspectives based on neutrality, good data, evidence, and numbers," was increasingly viewed as a setting in which smart, young staff were encouraged to pursue the transition from (in a telling semantic turn) "alternative" to "active" transportation.[7] This they did by passing a resolution encouraging the incorporation of Complete Streets into regional plans and policies in 2011; by commenting officially at meetings of the Planning Commission; by offering unofficial testimony as citizens; by highlighting in their planning documents the connections between transportation policy and climate change, health, and other arenas; and by, as "stewards, though not implementers, of our plan," educating other agencies within the region about innovations elsewhere. (All the same, a staffer, sardonically invoking the conviction that "if it works elsewhere, it is bound to fail in New Mexico," remarked that "We've learned not to talk about Denmark anymore.")

Nor did their work end at the region's edge. In a largely rural state in which policies to encourage walking, biking, and other forms of active living find little political traction, the council pressed the New Mexico Department of Transportation (which, an observer wryly observed, had evolved slowly and grudgingly from its earlier comfortable identity as a "highway department") to include in its own plan a section, drafted by the council, signifying that "We're part of a wider health promotion vehicle." (The section was briefly included and then excised.)

This lengthening parade of institutional advances drew energy not only from economic strains and electoral shifts but also from funding opportunities and legal requirements emanating from higher levels in the federal system. For instance, a staffer at MRCOG traced the agency's engagement with the health community to a joint commitment by the federal departments of Housing and Urban Development and Environmental Protection to "break down siloes" and create an expansive joint list of priorities, of which active living was one. Support from the National Park Service helped proponents of prescription (walking) trails to survey potential sites and launch their plans. Funds from the CDC enabled hospital representatives to partner with MRCOG in an inventory of walking trails and to conduct research on pedestrian safety. And, perhaps most important, a requirement of the Affordable Care Act that nonprofit hospitals conduct "community needs assessments" at three-year intervals, along with the sizable funds that accompanied the rule, elevated active living on the agendas of hospital chief executives, who were coming to value physical exercise as a noncontroversial means of health promotion they could comfortably embrace and who hired new staff to help them to comply. One such staffer recalled how "we partnered with the Community Health Councils to do our first needs assessment three years ago [in 2013] and came up with 'Community Health Priorities.' Active living was one of three."

The hospitals' needs assessments in turn brought them into closer contact with MRCOG. This connection, a hospital representative explained, "was a big gain for us because they see all that city, state, and federal stuff, and the expertise is there. There's a synergy. We use them and they inform us." In sum, the concerted forces of economy, polity, and federal activism shaped a policy environment in which active living initiatives not only addressed the needs of a widening range of stakeholders in the private, public, and nonprofit sectors of Albuquerque but also mapped a road to rewards for players who cooperated in the quest to advance those initiatives.

The city's evolving political economy not only encouraged public officials and private advocates who sought changes in the built environment in support of active living but also gave pause to actors in the bureaucracy and in industry who had stubbornly resisted such departures. A planner, whose appeals for innovation had long fallen on the deaf ears of engineers in city and county agencies, sketched in 2016 factors that had triggered

> a big change in the last five years, like day and night. Some of it is that federal money for road building projects to the suburbs is not here anymore. So if you look at the latest long-range transportation plan, instead of it being all about more and wider roads, more capacity, it's about a more diverse kind of transportation system, more about maintaining the existing system, not expanding it. Some of it is that we planners keep working at it, at building relations with engineers and getting them to listen. And some of it you see if you go to conferences and read professional publications. These trends are all around. Five years ago this job was totally frustrating. It was like planners and engineers were on totally different planets. It's so different now.

Shifts in federal transportation funding and in dialogue with and within the engineering profession both contributed to and were reinforced by passage of the Complete Streets ordinance. A legislative supporter of the measure opined that it "forces the issue with agencies, gives us a foot in the door. Yes, there's still resistance, they'll still cheat on the width of lanes and so on, but we can oversee them now. And the people who are sick of speeding in their neighborhoods are outnumbering the engineers." A city planner added that the law "means no more pleading and begging by us for engineers to do stuff. [Since Complete Streets passed] we've had more conversation and coordination with them than ever before. We said: 'We're gonna change your standards,' and [the head of a local agency] didn't fight. He sees the writing on the wall."

This writing, moreover, stretched into policy as Complete Streets and the proposed IDO supplemented (or supplanted) traditional engineering manuals with several alternatives authored by groups such as the National Association of City Transportation Officials. A planner on the staff of the city council explained that these alternative formulations "open up the framing of the issue and say you've gotta deal with multimodal options." In practice, "Even for resurfacing after the winter, when there are fifteen to twenty roads to be repaved, it says you can't just leave them as they were, you gotta put in bike lanes if that is called for, you gotta address intersection problems and modernize the roads and address calls for sidewalk improvements. Yes, they're expensive, and potholes drive demand, but Complete Streets does take account of sidewalks."

In 2016 active living initiatives in Albuquerque advanced, as they had in 2010, in a series of "yes, but . . ." steps. The "buts," however, had grown less inhibiting and the "yeses" had gained momentum. For example, prescription trails continued to get a lukewarm reception from busy physicians but were said to find favor with many nurses, nurse practitioners, office managers, midwives, and veterinarians—and also with an imposing new ally, hospitals that promoted walking in their community needs assessments. The implementation of Safe Routes to School remained sluggish, but the advocates for prescription trails had joined with the school officials and the National Park Service on a project that offered free pedometers, free park passes, and other inducements to encourage fourth graders in several schools to walk at least seventeen miles along a designated escarpment each school year. The city remained highly dependent on cars, but in 2015 "a coalition of public and private interests [came] together in Albuquerque to launch New Mexico's first bike-share program."[8]

Many neighborhood associations continued to wear self-protective "blinders" that, as a battle-weary planner put it, "make people think only of 'our' area and treat the rest as a black hole." But well-designed and -marketed visualizations of pedestrian- and bike-friendly projects were piquing interest. "We hear from the public, and they want walkability," insisted a planner on the staff of the city council. Demand, moreover, was shining a spotlight on supply.

> We're trying hard to push on zoning as it relates to the vertical built environment. The idea is to encourage people to park once and then walk to other vibrant activity centers. People like to walk and shop. For example, the ABQ Uptown [mall on Louisiana Boulevard] is a big deal here. It's great for pedestrians, and now there's demand for a downtown Main Street product like it. Walking in general is being sold as a tool of economic development.

Some counties in metropolitan Albuquerque continued to push for, as one advocate put it, jobs above all and "roads, road, roads." "It's the same old tension, progressives versus the small towns, where people say about walking and biking 'that's not our reality, we need to get places fast.'" But the decline of public funds to build new roads and the high cost of commuting on them curbed enthusiasm for sprawl. "It's still easier to go to the middle of nowhere and build cheap houses," a planner remarked, but mixed-use development was gaining ground in the city's downtown, and although developers of single-family homes continued to be the mainstay of a powerful industry and political lobby, observers noted that several developers had begun to specialize in multimodal construction, which was attracting the attention of their peers. "They see a project do well and then the idea gets comfortable for them."

Some engineers in the city and county stayed skeptical of and resistant to the new rules adopted in the Complete Streets legislation, but the momentum that had won unanimous support for the law in the city council, the marshaling of that support for efforts to enact an IDO, the updating of subdivision ordinances and the Development Process Manual, the increased activism of the regional planning agency, and adaptation and innovation within the engineering profession combined to increase both the formal levers and the practical leverage that advocates of active living could deploy.

In sum, although the diverse guardians of Albuquerque's built environment had not become fully fluent in the new vocabulary that planners had seen coming into vogue in 2010, in 2016 they appeared to be cautiously mastering its syntax and gradually embodying the signs, symbols, and substance of active living, and of a built environment hospitable to it, in a kind of second civic language that spoke, albeit softly, of a shifting cultural landscape.

4

SACRAMENTO

Active Living as a Breath of Fresh Air

Sacramento, a city of roughly half a million residents in a metropolitan area of 2.4 million, is the seat of Sacramento County and the capital of California. Founded during the gold rush of the 1840s, the city, which sits at the confluence of two rivers, became a convenient base for adventurers hunting gold in the Sierra Nevada and a hub for shipping and transportation. (It was, for example, the western terminus of the Pony Express and of the first Transcontinental Railroad.) Soon after California became a state in 1850 Sacramento was named its capital (1854), a designation made permanent in 1879.

Sacramento today has an economy dominated by government employment (mainly with the state, but also with the county, the city, and school districts) and health care institutions (branches of the University of California in nearby Davis, Kaiser Permanente, and Sutter Health Care, for instance). It is also a site of considerable social diversity: 33.6 percent non-Hispanic White, 28.1 percent Hispanic, 18.4 percent Asian, and 13.7 percent African American.[1]

STATE AND ENVIRONMENT

Sacramento is distinct among the five cities examined here not only because it alone is a state capitol but also because its pursuit of active living unfolds in a highly distinctive context—namely, the state of California's large portfolio of policies that aim to protect the natural environment (especially air quality) by influencing the character of the built environment. Across the roadways of the nation's most populous state (nearly 40 million residents) roll a superabundance of cars and trucks, which have gradually come to be recognized as a prime source of the air pollution that has troubled policymakers since the 1940s, when intense "smog" became a " fixture of life in Los Angeles" and led to the creation in 1945 of that city's Smoke and Fumes Commission, "the first of its kind in the United States and California's opening salvo in its fight to develop its regulative capacity to address air pollution."[2] When the federal government in 1970 passed ambitious legislation to protect the quality of the nation's environment (including, of course, the quality of its air), California stood ready with an even more stringent environmental protection statute of its own, the California Environmental Quality Act, which instructs state and local agencies to "disclose and evaluate the significant environmental impacts of proposed projects and adopt all feasible mitigation measures to reduce or eliminate those impacts."[3]

These assertive policies to protect the environment were, and continue to be, political products of a unifying "premise and matrix of political agreement," shared by governors as diverse as Pat Brown, Ronald Reagan, Arnold Schwarzenegger, and Jerry Brown, and dubbed by historian Kevin Starr the "Party of California." This distinctive political culture and context has powerfully shaped the state's policies on the environment (natural and built), transportation, and active living—and on the links among these domains.[4]

In the decades after the 1970s California worked to contain greenhouse gas (GHG) emissions with tenacity unequaled elsewhere in the federal system, and under Democratic and Republican governors alike.[5] This commitment intensified in the opening years of the twenty-first century. In 2006 the California Global Warming Solutions Act (AB32) set ambitious targets for the reduction of emissions, followed two years later by SB 375, the Sustainable Communities and Climate Protection Act, which focuses specifically on the roles of transportation and land-use policies in achieving those reductions. The law calls for large declines in the use of single-occupancy vehicles and requires that regions expand their cooperation in making decisions about transportation, land use, and housing; that they create "sustainable community strategies" to cut emissions; and that they include these strategies in their Regional Transportation Plans. In 2009 the state obtained from the Obama administration a waiver to enforce limits on GHG emissions stricter than those imposed by the federal Environmental Protection Agency.

In 2012 the state won federal approval to strengthen the emissions controls for which it had secured a waiver in 2009 and also began to implement a cap-and-trade system, a form of market-based regulation that aims to encourage investment in innovative clean technologies by allowing firms to trade permits to emit GHGs under a state-set cap, lowered annually. The following year the state's attack on pollution changed both managerially and methodologically. A new law merged several grant programs into a new Active Transportation Program in which counties and cities could compete for support. Meanwhile, SB 743 supplemented traditional measures of environmental impact, such as auto delays and level of service, with new criteria of evaluation, most notably vehicle miles traveled, or VMT, a more rigorous metric capturing the volume and length of trips, as a guide to balancing the control of traffic congestion with statewide goals that explicitly emphasized the promotion of active transportation as a means to combat GHG emissions.[6]

The scope of the state's environmental agenda continued to expand. In 2016 SB 1000 strengthened requirements that the general plans of counties and cities promote "environmental justice" by taking steps to protect disadvantaged citizens from toxins. And mid-2018 found the state Senate at work on AB 2434, which would establish a Strategic Growth Council and a Health in All Policies program, charged with expanding the prominence of health and equity in regional and local plans for transportation and land use.[7]

The component of active living called active transportation is not, of course, the sole or primary strategy for improving the quality of California's air. But even a cursory sketch of the evolution of the state's policies shows an intensifying focus on the contributions of both walking and biking to the reduction of GHG and air pollutant emissions and on the role of such reductions in enhancing the health of the public. A transit official in Sacramento put the point succinctly: "All these Assembly and Senate bills in our state really are a big push for active living because all the general plans have to address reductions in greenhouse gases, and that means not just new building types but also alternative transportation modes." In short, California's environmental protection regime constitutes an enabling context for the pursuit of active living in Sacramento that goes far beyond what one finds in the other four cities in this study. What entrepreneurial openings, then, did the city's active living advocates try to seize?

ALbD IN SACRAMENTO

When the Robert Wood Johnson Foundation created the Active Living by Design (ALbD) program in 2002, Sacramento was well acquainted with the need to reduce pollution (the Environmental Protection Agency had designated the metropolitan area, including Sacramento County, as a nonattainment area for ozone standards), was well versed in the arguments for active transportation, and was fairly adept at putting it into practice, especially by means of biking. Selection as one of twenty-five communities that won the grant confirmed and reinforced the aspirations of local actors who were eager to make the city and the unincorporated areas around it in the county still more receptive to active living.

One such actor is Anne Geraghty, who had trained as a city planner and in the late 1990s was working on transportation and land-use issues for the California Air Resources Board. Geraghty recalled her annoyance that a poorly designed intersection caused her to arrive late one day to a meeting of the air conservation committee of the local Lung Association and her consequent decision to do something to fix the problem and others like it around town. Inspired by WalkBoston, Geraghty founded WALKSacramento in 1998 as a voluntary organization with a mission to amplify the voice of the city's pedestrians amid and in collaboration with the city's many bicycle groups. "Our main focus isn't health," said an ALbD activist. "It's freedom—to walk."

When the Johnson Foundation called for proposals for ALbD, Geraghty and members and staff of WALKSacramento pondered whether to apply, what they might do, and who should do it. Having collaborated with the California Bicycle Coalition in search of funds to launch Safe Routes to School (SRS) projects in Natomas (a community near the Sacramento airport), Geraghty expanded that coalition in the quest for ALbD funds. When they won a grant, with WALKSacramento as grantee, Geraghty and colleagues pursued an agenda centered on the expansion of SRS projects, the elaboration and refinement of analyses to be deployed on behalf of the interests of pedestrians and cyclists in the land-use reviews performed at various levels of government, and the promotion of Complete Streets.

SAFE ROUTES TO SCHOOL

The conveyance of children by car and bus to and from school may not be a burning issue in the fight against GHG emissions, but it finds a place on most lists of challenges to active transportation and has the potential to affect the health of a sizable population—so much so that one planner opined that getting kids to walk or bike to school yields "the most bang for the buck" among active living strategies. As ALbD got under way in the mid-2000s, SRS, which won federal authorization in 2005 and aimed to encourage "a healthy and active lifestyle from an early age," had two conspicuous practical merits. First, it could be supported largely with federal funds, no small attraction for activists constrained by limits imposed on local resources by Prop 13 (a ballot initiative, approved by the state's voters in 1978, which capped annual increases in property taxes for homes, businesses, and farms), and not much eased by the modest sums proffered by ALbD or other grants. Second, SRS funds could be used for road improvements not merely "in" or "on" school grounds but also "in the vicinity" of participating schools.[8] This flexibility furnished benefits, as a former head of WALKSacramento explained, "beyond kids, to everyone walking there."

Institutional encumbrances counterbalanced these strategic appeals, however. SRS projects required "working with" the schools, a challenge that entailed protracted negotiations with power holders in a school system managed by an elected school board and an appointed superintendent who presided over multiple districts composed of several schools, each with its own principals and teachers. The formal independence of Sacramento's schools from county and city government meant that championship (or at least acquiescence) by the superintendent and (at least some members of) the school board was crucial to the progress of SRS. But support at the top by no means assured cooperation lower down the organization chart because, as one advocate put it, "schools are each in their own orbit, like little personalities all their own. Some work with us on SRS, some just want to be left alone." Another supporter cautioned that "with SRS you gotta gauge latent demand. It's not a Field of Dreams scenario, not just pouring concrete. If you build sidewalks kids won't necessarily start walking. You gotta encourage parents and school officials, gotta advocate and offer incentives." The political terrain reprises, mutatis mutandis, the chains of multiple clearance that bedeviled the workforce projects Jeffrey Pressman and Aaron Wildavsky studied in Oakland: unless school boards, superintendents, principals, teachers, and parents can be lined up in support of SRS, projects linger and languish.[9]

For these reasons, SRS is a quintessentially political exercise, one whose headway in Sacramento depended heavily on the championship of a local activist, Ted Link-Oberstar—cyclist, son of a former U.S. congressman, and parent eager to improve the schools his children attended—who pondered and interpreted the institutional context and set about building support for SRS. Link-Oberstar recognized that winning cooperation of the schools was above all a matter of reciprocity—the seeking and giving of advantages by those demanding and supplying change. In an interview in 2008 he explained the intricacies of the exercise:

> We were pondering how to get the School District to buy in. We did a big PR event—walks and a parade. I developed a resolution for the School Board, saying that it would be pleased to do SRS. Of course it was easy to say, but they had no time or money for it! Still, it got them on record on the matter. And I had helped to elect three school board members, which was a really good start, and we built on that. Over time, we had more events, and the ALbD grant, and people took co-ownership of it, which was key. And then the school board hired a superintendent who loved the idea. You can't plan that! And it's been good that I've been involved at different levels and in a range of issues, not just SRS. We have access because we're not just Johnny One-Notes, not just SRS, not just gimme, gimme. Because we're invested in a range of school issues—PTAs, tutoring, and more—we can get that letter of support. We've earned trust and credibility in the community over time. Ultimately, that makes it much easier to get that letter of support when you ask.

Although SRS projects in Sacramento gained momentum from political championship much stronger than that found in Albuquerque (discussed in the previous chapter), implementation was slow and uneven. Sources cited several familiar reasons why. Multitasking school principals were often loath to tackle new commitments bearing little overt relation to academic goals and tasks that were subjected to ever more exigent evaluation metrics. The enthusiasm of parents and school officials asked to guide the projects to completion was hard to sustain. Parental fear of "stranger danger" closed some parental minds to SRS a priori. Choices about use of the funds—for example, infrastructure versus demand-boosting public relations messages and events—could be divisive, all the more so once the fiscal context changed at both the federal and state levels. In the 2012 federal transportation reauthorization bill, Moving Ahead for Progress in the 21st Century (MAP-21), federal funds for SRS were merged with other programs in the new Transportation Alternatives Program. A year later California merged its Transportation Alternatives Program and related funds into the above-mentioned Active Transportation Program, in which communities competed for grants under various active transportation rubrics, including but not limited to SRS. These fiscal rearrangements, however, did little to settle disputes over how the money should be spent.

Persuading local voters to approve bond measures for objectives inclusive of SRS was one option, at least in some districts. But the best uses of such funds can be as contentious as their source. Proponents disagreed on the relative merits of infrastructure projects (bike lanes, sidewalks, and such) and those of non-infrastructure endeavors ("awareness events," for example). Or perhaps what one activist called an "in-between" approach—"money for what you might call people infrastructure to make things happen"—made more sense: "If you're gonna award money for SRS to school districts, you should view it as seed money to keep it going beyond the money itself. It's hard. The weakness is, a school district gets $100,000 to hire someone, then the money goes, the position goes, and what's left? You can use a grant to get it going, but the *community* needs to be able and willing to make it durable." By 2018 SRS had taken root in a few

Sacramento schools, but the "bang for the buck" the effort yielded was unknown, and perhaps unknowable.

WALKING

Sidewalks are to pedestrians as bike lanes are to cyclists—a key object of attention, analysis, and advocacy—and the promotion of better sidewalk standards for the city and county was a high priority for WALKSacramento from its outset. Policies for sidewalks, however, involve much more than pouring concrete in the right places, and they are, moreover, only one (albeit a central) element in a pedestrian-friendly built environment.

Sources interviewed in 2018 typically prefaced their comments by remarking that not so long ago—as recently as a decade—the priorities of pedestrians were, state directives notwithstanding, little more than an afterthought in regional and local land-use and transportation planning, which at best grudgingly "accommodated" them at the margins of decisions. Interviewees generally concurred, however, that the picture had changed dramatically. The benefits of active transportation, including walking of course, claimed a larger place in the spotlight as the state intensified its pressure on regions and localities to reduce emissions of GHGs. Implementation of the national Americans with Disabilities Act and the risk of suits against jurisdictions that failed to comply with its requirements concentrated the minds of planners on issues such as the design of curbs, which affected access not only for the disabled but also for pedestrians in general.

At the regional level, Sacramento's Council of Governments (SACOG), the designated metropolitan planning organization, was constrained by its heterogeneous composition and the pro-growth sensibilities of many of the county and city officials who served on it. Nonetheless the council was assertive in seeking to bring patterns of land use and development more closely into line with the state's environmental targets and with its expectations for walking and biking as favored means to meeting them.

In 2002 the state's Department of Transportation in the administration of Governor Gray Davis had issued a "blueprint" replete with "goals and performance measures" to promote biking and walking.[10] Then, with the help of a smart growth leader recruited from Portland, Oregon, SACOG proceeded to develop in 2004 its own Sacramento Region Blueprint, which included, among many smart growth principles it sought to promote, "options for people to walk, bike, or take public transportation to work and play."[11]

The development of the Blueprint, a thirty-month process that included a vivid visualization of growth-as-usual compared with scenarios based on smart growth principles, broad public involvement, interactive planning exercises, and an uncommon attention to the prospects for implementation (lest the exercise become yet another sad example of "stranded inspiration"), may indeed merit its self-conferred designation as a "New Paradigm for Transportation Planning."[12] And in subsequent years the SACOG never wavered in its strong advocacy for pedestrian-friendly policies. In 2016, for example, its metropolitan transportation plan endorsed, among other active-living-friendly measures, "sidewalk gap closing, [Americans with Disabilities Act] retrofits . . . intersectoral improvements and more Complete Streets connections."[13]

This growing solicitude for active living also infused the intermittently updated general and master plans that governed, at least in principle, decisions by the county and city about the built environment. These provisions were far from perfunctory. The county's general plan, adopted in 2011 and intended to guide development through 2030, includes a 172-page pedestrian master plan that addresses (among other topics)

collisions between pedestrians and cars (39–41);
the lack of sidewalks (42);
the need for curb ramps, cross walks, crossing islands, and pedestrian-oriented signals as well as sight distances and the timing of signals (42–44);
buffers, trees within buffers, the encouragement of low-traffic volumes and speeds, and the appropriate width of sidewalks (44–45);
how to meet the needs of the blind and wheelchair-bound even in rural areas where sidewalks are not wanted and pedestrian overpasses and pathways are often abandoned for fear of crime (45);
the need for maintenance of all of the above (48–49);
extra safety measures on routes with big trucks (49);
the need to fix broken sidewalks and those with gaps, to improve lighting, and to install audible pedestrian signals (69);
coordination with bike plans (72);
school crossing guards (76);
methods of traffic calming (radar trailers, message sign boards indicating vehicle speed, speed bumps, additional stop signs) (77);
utility poles and street furniture that obstruct walking (79);
the addition to streets of public art, landscaping, and resting benches (83);
the modification of zoning codes and development standards (86);
yield-to-pedestrian signs (113);
way-finding signs, fencing, and call boxes on pathways (133); and
how to "market" walking by means of "a variety of media and events" (135).[14]

All of this, moreover, was to proceed within salient constraints—namely, figuring out how to pay for it, subjecting plans and projects to public input, and assessing both the cost-effectiveness of proposed improvements and the demands of equity in their location (pp. 150–63).

Because the practical implications of these aspirations are never self-evident, they must be constructed in detail, case by case, and in a variety of contexts, not all of which can be called enabling. For one thing, it can be cumbersome and costly to retrofit updated sidewalks into communities built up years ago. Pedestrian advocates are perpetually looking for targets of opportunity, but a good walking environment is more than the simple sum of accessible blocks—respondents frequently invoked the term "connectivity" to indicate the importance of an unimpeded flow of movement.

Sidewalks—and, more important, an appealing network of well-connected sidewalks—are most easily introduced into designs for emerging residential, commercial, and public projects that are not yet literally set in concrete. For example, a city official recounted how a development project in Natomas, an unincorporated area to the north of the city in Sacramento, was

> going great. There's real density out there. But at fifteen homes per acre there's no room for shade trees, so we rewrote the sidewalk and street specifications. Sidewalks must be a minimum of eleven feet wide. We mandated minimum six-foot planting strips, and then the sidewalks, then the front door. This way you lose one in seventeen homes. It's good for getting people out of that car.

What pedestrians regard as optimal sidewalk policy may not prevail so readily over the preferences of private developers, however. A county transportation planner remarked that state law benefits the environment, but "the trade-off is all the studies that are required, which grow constantly and which cause delays and drive up the cost of projects." The "fiscalization" of land-use policy, a result of limitations that Prop 13 imposes on local property tax levels, puts a premium on acquiring deep-pocketed taxpayers such as big box commercial centers and ensures that when developers threaten to take their business elsewhere, they get a respectful hearing from local land-use officials. The welcome extended to developers grew notably warmer when the recession of 2008 and beyond slowed new building severely, leaving officials reluctant to discourage, still less reject, projects that were (as one planner said plaintively) "all we've got."

Nor does the private sector hold a monopoly on resistance to pedestrian- (and cyclist-) friendly policies. A city planner recounted the frustrations of dealing with public agencies and public landowners:

> It's all a matter of leverage and we can "hold hostage" the private developers. But with state land, parks, and Caltrans [California Department of Transportation] it's much harder. Caltrans's mission is not ours. Same with the railroads—all they care about is moving freight. Their headquarters is in Omaha; it's a different world. They own lots of significant land, but we can't work with them. We can't get over- or underpasses to connect two sides of the track! So you end up with very tall bridges just to go over a train, which is crazy. Railroads are a big barrier in California. If a project involves railroads, you're dealing with the Public Utilities Commission, and it drags on and on.

Finally, attempts to make the built environment more hospitable for walking may meet objections within the communities the improvements are intended to benefit. In 2018 a prominent proponent of active living cited as a case in point a proposal to build a car wash across the street from a light rail station in a suburb of South Sacramento:

> The planners recommended that it be denied as contrary to the zoning code and county plans—we didn't want more cars in that area. We had on our side SACOG, regional transit officials, and WALKSacramento. But the developer touted it as a source of economic development even though it would provide jobs for only three or four minimum-wage workers. The developer lobbied the commissioners and brought in community members, all African American, who *did* speak for the car wash. They made a good point: there was already a lot of traffic near the transit station there—six lanes of highway, and it wasn't good to walk or bike there. We said, we could build a pedestrian bridge and so on, but the truth is they were right about how it is now. The commissioners said it wasn't right to overrule the desires of the community, and we lost big.

Asked whether this case is typical, the advocate explained the costs of political accommodation:

> There is always opposition, and it is widespread. You write plans for connected walking and biking areas, and then the city allows drive-through restaurants near a major transit stop, so why not a car wash as well? The commissioners are not going to be pedestrian friendly if you've chipped away so much of your original plan! If the developers and communities are mobilized, they *will* tend to prevail. Jobs and economic development are buzzwords.

Although Sacramento County and City have embraced plans that are "general" and presumably equipped to "master" changes in the built environment, a wide gulf looms between the formulation and implementation of policy. Plans get applied (enforced is too strong a word) through a meandering series of ad hoc decisions about particular cases, which active living proponents seek to influence by means of three main strategies.

First, over the course of two decades, WALKSacramento has developed a reputation for a "distinctive competence" (Philip Selznick's term) in analyzing the implications of land-use decisions for pedestrians. Capacity building began with the ALbD grant, which, an activist in WALKSacramento recalled,

> got us into land use issues. Before, we lacked resources, but then, with the extra time and staff the grant made possible, we got in early to influence decisions of the city planning commission about accessibility for pedestrians and bikes in private projects. And the local air district relies on us to comment on pedestrian amenities in development projects that are under review for their environmental effects and the options of alternative transportation.

By 2017 this role had been secured and refined. In that year, for example, WALKSacramento transmitted to the city's Community Development Department a four-page, single-spaced review of a proposal to build 293 multifamily apartment units on 10.3 acres in Natomas. Contending that the plan failed to conform to the General Plan and the Multi-Family Residential Design Principles, the analysts warned that it needed "drastic changes" and offered specific recommendations, including "placement of pedestrian access gates around the site," putting a "detached pedestrian path along Arena Blvd. that is separated from the roadway by landscaping and is well shaded and lit," and a range of other pedestrian-friendly modifications.[15]

Second, pedestrian advocates cultivated alliances with like-minded organizations with an eye on the built environment. A leading case in point is Design 4 Active Sacramento (D4AS), a team founded in 2013 by participants in a CDC-sponsored national public health leadership program. Determined to apply in Sacramento the Active Design Guidelines ("a manual of strategies for healthier buildings, streets, and urban space") crafted in 2010 by several public agencies in New York City and the American Institute of Architects, the group sought (among other goals) to increase the number of Sacramento streets suitable for walking and biking and to encourage mixed land uses with walkable districts.[16] It continued its push for healthy communities in reviews of the county's design guidelines and secured the adoption of public health, safety, and "livability" goals in an updated version of the Housing Element in the county's general plan.

In 2014 D4AS affiliated with WALKSacramento, to which its eleven members, led by the county's principal planner and sustainability manager, brought "even more expertise, diverse talents, and a shared passion for healthy built environments." When the county began updating its zoning code (which "arguably wields the most influence over the building and planning landscape") for the first time in thirty years, D4AS won endorsement of healthy communities and active design precepts in the guidelines adopted in 2015.[17]

The domain of "design" continued to expand. The city of West Sacramento, for example, participated in a Crime Prevention Through Environmental Design strategy, which worked to include criteria such as "natural surveillance" (high visibility of comings and goings) and "natural access control" (careful placement of "entrances, exits, fencing, landscaping, and lighting") to prevent crime, which is a prime deterrent to active living.[18] Sacramento County's public health officer, a founder of the D4AS team, also worked with eighteen recreation and park agencies on a program that gave residents written "prescriptions" to spend thirty minutes five days a week in local parks "walking or participating in an activity of their choice."[19]

Third, these allies pursued a bottom-up approach to building political support. In 2008 an activist explained how WALKSacramento and its allies worked to secure changes in the transportation components of the county's general plan: "In the past, traffic engineers and sometimes developers opposed us. This time we worked as hard as ever to give a rational basis for our ideas, but we also worked with the new head of the county [Department of Transportation], who's sweet and reasonable, and we talked separately with four of the five planning commissioners." Ten years later, these political skills were intact—and sharpened. An active living leader described how he "sits on the design review committee and asks a lot of questions about how a planned development fits with [land-use] studies." But, he continued, queries are one thing; influence is another. "We start at the bottom, with the engineers and planners, and work *up* to the pols, not vice versa. Our community development people talk to the builders right at the start. All these little projects add up to our built environment, after all."

The networks of and around pedestrian advocates have assuredly gained legitimacy, adroitness, and influence over time, but the value of the political capital thus amassed is difficult to gauge until it is tested in the course of bargaining over the design of projects and programs, one by one. An activist summed up the strategy: "Put it in the plan and good luck."

BIKING

Bike lanes are to cyclists what sidewalks are to pedestrians: a central but far from exclusive object of policy interest. Cyclists typically work to expand the number of bike lanes available in their jurisdiction, to be sure, but they also promote the addition of bike racks to mass transit vehicles, convenient facilities for parking bikes in commercial and residential buildings, bike-sharing programs with ample and well-located sites, and strict enforcement of laws against drivers who endanger bike riders. And just as the expansion of sidewalks contains multitudes of correlative objectives (green medians between sidewalk and street, for instance), so too do proponents of cycling push for bike lanes that are wide, clearly demarcated, protected from cars and trucks, and well-connected.

In Sacramento, as in California generally, the promotion of cycling, as of walking, has steadily gained prominence in state laws and regulations, in the documents that guide the work of the regional SACOG, and in distinct cycling master plans within the general plans of the county and city. The Bikeway Master Plan adopted by Sacramento County in April 2011, for example, is compendious indeed—344 pages, twice those of its pedestrian counterpart.[20]

The plan begins by lauding biking as a "low-cost, quiet, non-polluting, sustainable, and healthy" form of transportation (1) and declares that it aimed to make the experience "more safe, comfortable, convenient, and enjoyable" for cyclists. Drawing on a general advisory team, a technical advisory committee, public workshops, and surveys of public opinion (2–4), it invokes three "classes" of bikeways defined in state law (paths, lanes, and routes; 6) and then addresses a long list of critical issues. These include

parking (8, 164–68);
striping ("a white lane, six inches wide for bicycle lanes, and four inches wide for multipurpose lanes") (26);

consistency with the pedestrian master plan (28);
ease of access to light rail (29–31);
connectivity (the desire for "long uninterrupted rides") (50);
the need for better maps, signs, lighting, rest stops, and overcrossings (58–59, 91);
maintenance and security (162–63);
ways to augment demand for cycling by means of "encouragement, education, and enforcement" (179–198); and
Safe Routes to School (192–94).

The plan proclaims four explicit goals—an increase in bike use, a reduction in collisions by and injuries to bikers, growth in the number of bicycle facilities, and dedicated funding "proportional to mode share" for county facilities, transportation programs, and staff support (44–48). It cites problems and names names (for example, Riverside Boulevard was plagued by the "narrow width and poor condition" of its bike lanes; 41–42). Having sketched criteria to guide selection among the three classes of bike lanes for proposed projects in given areas (93), the document expatiates in appendices on priorities, timelines, and funding plans (appendix A); best planning practices observed in cities such as Portland, Oregon (appendix B); bicycle goals from other jurisdictions (cities, counties, states, Copenhagen, Vancouver) (appendix C); funding sources (appendix H); and a dozen or so endangered plants and creatures that must not be disturbed by new construction (elderberry plants and shrubs, tricolored blackbirds, California Tiger Salamanders, Giant Garter snakes, and native trees, to name a few; appendix I).

The plan supplied abundant inspiration and instruction for proponents of cycling, but, as with the priorities of pedestrians, those of cyclists are realized in practice largely by means of detailed and protracted skirmishes, waged by cycling advocates and their allies, over the application by city planners and the city planning commission of general "authoritative" language to particular projects. Cycling proponents may, for example, identify streets that seem especially well suited for the addition, expansion, or improvement of bike lanes and then pressure city or county authorities to take action by imposing a "road diet" that removes one lane from cars and earmarks it for cyclists. In new commercial and residential developments or in ones undergoing renovation, cycling advocates try to intervene early in hopes of securing bike lanes and other amenities (such as adequate storage rooms for bikes) before plans take firm shape.

In the pursuit of these initiatives, California's distinctive policy context confers important political resources. A veteran of the Sacramento Metropolitan Air Quality Management District noted:

> By linking the issues of development and pollution, state law gives leverage to nonpolluters like cyclists and pedestrians because it says the lead agency must disclose the environmental effects of a plan, which gives us leverage. We look at air quality analyses and ask does the development in question encourage walking and biking. And we rely on WALKSacramento and bike groups to comment on pedestrian and cycling amenities.

Advocates also argue for improvements in parks and other green spaces—for example, paths that are wide enough to accommodate bikers as well as walkers and are paved to provide a smooth, safe ride. Advancing the cyclists' agenda thus demands both constant vigilance on several fronts in order to identify targets of strategic opportunity and persistence in pressuring decision makers to give serious consideration to their requests.

Interviewees in Sacramento all concurred that "great progress" (in the recurrent phrase) had rewarded the labors of cycling advocates. A county planner cited one of many points of light: "We put a bike path to Folsom along the river and now it's *packed*." And the pace and scope of innovation seem to be growing. For instance, in 2017 the city launched Vision Zero, which resolves to improve five corridors with the highest number of serious and fatal crashes involving pedestrians, bikes, and motorists.[21] A year later the city began installing "parking-protected bikeways" (in which parking is moved away from the curb, creating a buffer for cyclists between cars and curb) in several sections of the downtown. May 2018 saw the arrival of a bike-share program in which a private company rents electric pedal-assisted cycles.[22] These accumulating programs would seem to suggest that, in Sacramento, active living centered alliances may be acquiring countervailing political power sufficient to reach a tipping point in policies for transportation and the built environment.

For cyclists, as for pedestrians, however, progress is rarely attained without surmounting economic, organizational, and political barriers. In Sacramento, as in the other cities studied here, putting a multilane road on a diet involves some financial cost and evokes opposition by (literally) path-dependent commuters whose coming and going will take longer. No one disputes the virtue of negotiations that "bring the residents along," but doing so is time-consuming, labor intensive, and iffy. On the other hand, taking unilateral action violates norms of consultation—"No one asked *us*."

To be sure, the incumbents of erstwhile highway departments have come to accept their new identity as "transportation departments" and the training of traffic engineers has evolved from grudging tolerance of nonmotorized conveyances to acknowledgment of the claims of multimodal and even active transportation. But active living proponents continue to contend that when tradeoffs arise in the course of implementation, the imperatives of safe and efficient "levels of service" for cars and trucks still rate higher with engineers than does ease of access for bikers and pedestrians. And these tradeoffs pop up in unexpected venues—for instance, in tensions not only with engineers but also with managers of the mass transit systems (with whom bikers and walkers are presumably allied in the quest for cleaner air) who complain that the region's large biking community overwhelms the capacity of their trains and buses to supply racks for their use.

In principle, the City's Parks Department and the County's Department of Regional Parks might be expected to open their arms to the cycling constituency. In practice, however, the park authorities were said to be preoccupied with budget shortfalls, struggling to maintain their parks, and pondering how to prevent them from becoming "gang havens" (as one source called them). A planner explained: "With the parks, you've gotta get involved early in site planning, which involves the parks and rec people, not city planning. Parks and Rec holds workshops with the neighborhoods, but it's hard to get formal notice of them." And anyway, said one active living aficionado, Sacramento tends to emphasize "green spaces" all around town rather than parks per se.

Finally, bikers no less than walkers are frustrated by the fiscalization of land use that drives local authorities to cater to the wishes of big potential taxpayers (and federal road builders), for whom sidewalks and bike lanes are an afterthought, if not a nuisance. Notwithstanding the ground they have gained, active living supporters complain that land-use decisions too often proceed in piecemeal fashion and give uncritical approval to development projects. "The idea is do it fast, instead of do it right," said one critic, who cited as an example a development that was "badly designed but justified by its being on a future rail line to the airport. They needed to approve it to get federal money, so they rushed it through. SACOG was not supportive, and some other groups too said it was premature. But the feds really wanted to fund that transit project and the county supervisors really wanted that federal money." A county planner lamented that haste made waste: "We aim for connectivity, but timing can defeat it. That 'one spot' can discourage walkers and bikers, and that 'missing segment' is especially important with SRS."

The vagaries of political economy, moreover, can produce questionable omissions as well as commissions. A Sacramento official observed that the suburb of North Natomas was, on the whole, receptive to active living, but a major highway divided a park in the northern part and bike trails in the southern sections of town. "We were going to build a bridge—a $6 million project, no cars on it—to connect them, and we tried to explain it on Channel 13. Their take on it: 'A $6 million bridge in a $59 million city deficit?!' I explained that it was cheap and used federal grant money besides. But on TV they portrayed it as the 'pricey bridge' that the city doesn't need."

COMPLETE STREETS

The biggest challenge for active transportation, a planner contended, is "linking bikes, pedestrians, and transit to each other and over different areas." This integrative task highlights the appeal of Complete Streets. Central to the ethos of smart growth, Complete Streets was a top priority of WALKSacramento from its start in 1998 when, an activist recalled, it "was just beginning to come on the radar screen." Over the next two decades, a county planner explained, Complete Streets steadily gained ground:

> Land use was not seen as a significant component of transportation till very recently. It used to be that bikes and pedestrians were just "accommodated." Now they're major modes. Complete Streets is a SACOG priority because we've got to support shorter trips, and its regional blueprint forced local governments to rethink and update their long-range transportation plans. Complete Streets is integrated into our general plan. The governor [in 2008] signed a Complete Streets bill, though without money attached, with a whole philosophy about land use linkage. Congressman [Bob] Matsui has a national Complete Streets bill. Even the mindset in Caltrans has changed.

Active living supporters recounted battles to capitalize on the newfound prominence of Complete Streets in the general plans of the city and county—battles that often found them opposed by transportation staff who objected to measures that would reduce local capacity to cater to cars. The pattern, similar to that traced by pedestrian improvements and bike lanes, was essentially one step forward, one back. A planner declared: "Smart Growth is part of the climate action plan in many California cities and Complete Streets is gaining ground. Developers are getting with it." Then he added the ubiquitous "but": "Mitigation [of emissions] clashes with the localities' revenue goals. Big box projects give you the tax dollars, and so projects get approved piecemeal, which is a problem for connectivity and Complete Streets."

Some interviewees voiced cautious optimism on the political front: in 2016 the city elected a new mayor, Darrell

Steinberg, who as a prominent state legislator had made much of sustainability and livable communities and as mayor worked to win the adoption by the city council of a formal Complete Streets policy. Observers conjectured too that Complete Streets would benefit from rising interest in infill development downtown, where a new basketball stadium was under construction, and the project seemed to be gaining residential appeal for millennials and for baby boomers planning to age in place. A staffer at SACOG described the "transformation of downtown since 2005" as "really dramatic." And although retrofitting sprawling suburbs to incorporate Complete Streets was no easy task, in 2018 the SACOG had convened a "corridor working group" and hired a consultant to explore opportunities to do so.

As usual, much depends on the particulars—above all, the distinct spatial character of communities and their willingness to accept the enhanced mingling of different social types that accompanies Complete Streets. A county planner contended that the hardest part of a move toward Complete Streets is "changing the status quo in existing communities. The NIMBY types will shut it down 'cause 'we got ours.'" A SACOG staffer opined that "communities of 1950s vintage are not entirely car-centric, and you could retrofit them toward Complete Streets. But those engineered in the '70s, '80s, and '90s were engineered *big*, the parcels are *big*, and it's harder." Installing Complete Streets in a general plan ("thirty thousand feet up," as a county planner put it) is one thing. Getting it incorporated in local zoning codes is another. Working within those codes to implement the myriad connections among streets, residential and commercial developments, and transit modes (and among the agencies and actors that shape these connections) that Complete Streets demands is something else again.

Nor is it easy to evaluate the effect of Complete Streets policies on variables such as the safety of pedestrians and cyclists. For example, a study addressing that question in Sacramento lamented a lack of data that "hampers researchers' ability to analyze the bicycle and pedestrian safety environment and the impact of interventions like complete streets"—a gap that stymied interpretation of the results of an investigation of three such projects, two of which saw reduced crash rates but one of which recorded an increase.[23]

The economics of housing further complicate the quest for Complete Streets, as for other active-living-friendly elements in the built environment. Having weathered a near-collapse of new housing starts during the recession that began in 2008, Sacramento grappled with an "affordability crisis" that afflicted both the city's middle class and its (many) new arrivals and pervaded debates about land use and development throughout California in the second half of the 2010 decade. Carol Galante, a housing researcher, explained: "We've had a huge increase in population and a huge increase in jobs, and we do not have anywhere close to the supply of housing to put people."[24] As a SACOG staffer put it, "It's a messed up market with low inventory and high costs. [In such conditions] active living goes down on the Maslow hierarchy." Meanwhile, sprawl, a county planner conceded, "is still the cheapest model." Even in progressive Sacramento, therefore, the integrative vision of Complete Streets could not escape the durable realities of fragmentation. "Holistic" has a pleasant ring, but, as one source remarked, "You gotta pick and choose your interventions. We didn't choose the smartest growth, but we chose as smart as we could."

The tenacious commitment of policy leaders in California to the reduction of air pollution has created an enabling context for active transportation, and thence for active living, stronger than that of any other American state. If agenda setting and the formulation of public policy were the main explanations for policy outcomes, Sacramento, working within that bracing state context, would presumably behave differently from, and achieve results superior to, the other cities studied here, operating as they do in state policy settings less determined to prod the built environment toward the goals of active living.

In fact, however, the active living policy process in Sacramento bears a striking family resemblance to that in Louisville, Wilkes-Barre, Albuquerque, and New York. To be sure, the process in Sacramento is shaped by state laws, rules, guidelines, and communications to which attention must be paid in regional blueprints, county and city general plans, and pedestrian and cycling master plans. But the decisions that shape and reshape the built environment are highly contingent: on local particulars, the most important of which is the intensity of support or opposition to specific proposals in distinct sections of town or county; on the willingness of officials in a range of organizations—the Departments of Transportation, Public Works, and Housing, the schools and more—to cooperate in putting proposals into practice; and on the energetic championship of local leaders in the public or private sectors on behalf of active living initiatives.

In Sacramento (as elsewhere) aspiring policy entrepreneurs canvass environments—built and political—on the alert for openings (grants, public requirements that demand compliance, economic gains for developers, enthusiasm among well-organized advocates, and so on) and then place their bets and try their luck. This educated guesswork is often framed in the integrative and holistic lingo of smart growth, New Urbanism, Complete Streets, connectivity, and sustainable communities, but—constrained as it is by settled meanings attached to place, by disparate missions and motives in the organizations whose support is vital if projects are to advance, and by the reluctance of leaders to spend political capital on initiatives that may generate more conflict than reward—active living advocates play a piecemeal game in hopes that their fragmentary victories will maintain momentum for more triumphs and in time add up to a built environment in which well-designed connections will more reliably encourage physical activity.

In Sacramento, as in the other sites, attempts to tailor the built environment more closely to the requisites of active living have confronted both the cultural entrenchment of forces favoring sprawl and the raw power of road builders, developers, and other interests that benefit from the continued infusion of funds into traditional patterns of metropolitan development and, no less important, insist on the need to keep the state's vast system of roads in decent repair. (In 2017 California passed SB1, the Road Repair and Accountability Act, and created new transportation-related taxes and fees for that purpose.) Within that durable context, the activists' game is an exercise in countervailing power—in building support in affected communities, in discerning common ground at critical levels within and between organizations, and in identifying and helping to forge political rewards that outweigh perceived costs in the eyes of potential supporters. The consensus among local players and observers that in the decade between 2008 and 2018 Sacramento made long strides toward a built environment that supports active living suggests that the advocates have learned how to play the political game from formulation to implementation of policy with considerable speed and skill.

5

NEW YORK CITY

Flourishing at the Margins of Policy

At first glance, New York City would seem to be an inhospitable venue for active living initiatives. The city is choked with traffic, and its pedestrians are said to come in two varieties: the quick and the dead. Its physiognomy has been decisively shaped by the massive highway-building projects of Robert Moses, the master builder of Robert Caro's celebrated biography. And the city's late-twentieth-century mayoral leadership had many more pressing priorities—for instance, fighting crime, the top item on the agenda of Rudy Giuliani, mayor from 1994 to 2001—than the promotion of walking, biking, and other forms of physical exertion.

First impressions can mislead, however. Under both Michael Bloomberg, mayor from 2002 to 2013, and his successor, Bill de Blasio, the city launched several impressive programs that aimed to make walking and biking safer and more attractive and to augment physical activity in its schools and parks. This chapter explores the sources and outcomes of this civic activism.

PUBLIC HEALTH IN CITY HALL

That active living should find a niche on the agenda of Mayor Michael Bloomberg is unsurprising. Bloomberg believed that strong municipal leadership could do much to improve the health of local populations and (as demonstrated by the $500 million gift he had given to the School of Public Health, now named for him, at his alma mater, Johns Hopkins University) he recognized that public-health programs were central to the pursuit of that goal. Wealthy, widely traveled, and cosmopolitan, Bloomberg was well informed about the practices of other "global" cities such as London (where he had a home) and of international exemplars of best metropolitan practices. An official in the city's Department of Health and Mental Hygiene (DOHMH) recalled that the mayor "saw the transportation angle—biking, walking, congestion pricing, and so on—as part and parcel of public health. He was interested in Copenhagen, which was doing great things with bikes. Active living was also a way to deal with climate change—healthy for you *and* good for the environment."

Like many other mayors (including those discussed in previous chapters), Bloomberg wanted his city to excel in competition for smart, young executive types, a growing number of whom were said to value walkability and the opportunity to make do without a car as part of an appealing urban lifestyle. And although few rank-and-file New Yorkers routinely ride a bicycle amid the city's "canyons of steel" (in the words of a song by Vernon Duke), many cover considerable ground on foot in the course of daily life, as do straphangers as they access and move about within the subway system. A prominent slice of elite opinion in New York insists that Jane Jacobs (iconic denizen of Greenwich Village, celebrator of urban density, and author of *The*

Death and Life of Great American Cities [1961]) may have lost some battles with Robert Moses in the postwar years but has in time won the war to shape the built environment. Moreover, much of the city's disproportionately liberal and Democratic electorate responds favorably to "causes" such as sustainability, greenness, New Urbanism, and smart growth, to which active living is generally viewed as complementary, even integral.

Bloomberg, moreover, not only "got" public health but also put it at the top of his mayoral agenda, remarking that the measure of achievement that meant most to him was the rise of life expectancy in the city's population. The mayor was committed to "doing something" about problems, not "wringing your hands"; wore a thick skin against bad publicity; and enjoyed a large personal fortune that gave him freedom from the need to solicit campaign contributions and into which he was prepared to dip to fight back against well-heeled obstructive groups. Armed with firm views about the vast potential of public health to prolong and improve lives in and beyond New York and eager to lead the profession from its traditional preoccupations into an assault on chronic diseases by means of an evidence-based commitment to prevention and health promotion, Bloomberg was increasingly recognized as "the nation's loudest public health advocate"—indeed, as its "biggest voice in health."[1]

The political prominence of public health in city hall created an institutional context in which the DOHMH was invited—indeed, expected—to craft policies that would demonstrably reduce mortality and morbidity within the five boroughs and could count on the support of top-level champions when the going got tough. Early successes in campaigns against smoking and unhealthy foods had, by 2007, won admiring national attention for the department, and the sense that "exciting things were happening" in it attracted talented people from a range of disciplines and backgrounds to this "powerhouse of a health department . . . that broke news and set a national agenda in public health."[2] The agency's mission and national visibility, in short, extended an open invitation to public-health policy entrepreneurship.

These contextual considerations offered potentially fertile ground for active living initiatives, but their emergence was by no means automatic. Active living was not part of the core missions of the Departments of Transportation, City Planning, Design and Construction, Parks and Recreation, or Education, but "even a relatively minor public undertaking" may involve these and other agencies—"each with its own procedures, paperwork, and goals."[3] And although active living fell within the portfolio of the DOHMH, that capacious collection contained many other health-enhancing strategies, such as antismoking and healthy eating initiatives, with which active living competed for attention and resources. It competed, moreover, under a strategic disadvantage: the laws, regulations, and taxes that could be marshaled against smoking and unhealthy foods were applicable uneasily if at all in a public push to discourage physical inactivity and sedentary lifestyles.

Signs and omens in the early Bloomberg years did not seem to favor active living. The commissioner of the Department of Transportation, a holdover from the Giuliani years, was at odds with Transportation Alternatives, the city's leading advocacy group for pedestrians and cyclists. Health Commissioner Thomas Frieden outlined ten "priority areas," such as "be tobacco free" and "get checked for cancer," and although two goals ("keep your heart healthy" and "have a healthy baby") might be advanced by physical activity, the agenda made no explicit mention of active living or the built environment.[4] A review of the city's health activities between 2002 and 2007 had little to say on these matters, referring only to "physical activity programs" and "assessing fitness" as initiatives expected to promote child health.[5] In New York, as in many other communities, active living was on the municipal agenda but not high, still less uppermost, on the agendas of leaders of the Department of Transportation (DOT), of DOHMH, or indeed of any particular government official who enjoyed authority or influence.

Agenda-setting for active living got a boost in 2004 when Lynn Silver became assistant commissioner for chronic disease prevention and control in the DOHMH. Silver had recently perused research funded by the Active Living Research arm of the Robert Wood Johnson Foundation's Active Living by Design program (published in the *American Journal of Preventive Medicine* in 2005) and had been, as she recalled several years later, "impressed by [that research] for a decade." Surveying the scene at DOHMH, Silver found within the department some projects on worksite wellness and made plans to use the city's regulatory authority to require more play activities for kids in day care. But how to gain leverage on the built environment, for which policy responsibility sprawled across local (and state and federal) agencies, was not obvious.

Serendipitously, family connections kicked in. Urged by her cousin, an architect, to seek contacts within that profession, Silver discussed strategy with the wife of Rick Bell, president of the American Institute of Architects (AIA). Bell agreed to collaborate with DOHMH and to convene a "Fit City" conference, to which architects, planners, and officials of city agencies were invited, to discuss how new approaches to design of the environment might make it more conducive to active living. The first year of the collaboration, Silver recalled, was "hard, and packed, but fun. The meeting was successful, the visual thinking of the architects was great, but buy-in by the city agencies was low. We were not yet getting the public sector colleagues we would need."

The institutional prospects for collaboration brightened when in 2006 Silver met Karen Lee, a physician and public-health professional then working at the CDC on the prevention of chronic disease, who had delved deeply into the preventive properties of active living and the importance of the built environment in promoting it. Eager to elevate active living on the agenda of DOHMH, Silver hired Lee, who proved to be an energetic built-environment director and coordinator for active living initiatives in the department, encouraging the creation of a built-environment team to work on the issue and seeking to mobilize foundation support for their plans.

Lee pictured her role as that of a full-time active living specialist who framed issues in ways that elicited interorganizational and intersectoral cooperation. In an interview she explained that "you need a dedicated person to rally across the different departments. It has to be somebody's *job*, and the full-time aspect matters because it signals that it is important." Convinced that "the critical thing is structuring conversations, and how you do it," Lee began convening workshops and meetings at which representatives of a range of city agencies, organizations (such as Transportation Alternatives), and professional bodies (architects, for example) discussed the case for an active-living-friendly built environment, the evidence that policies seeking to promote active living yielded demonstrable gains for health and other social objectives, examples of best practices in other locales, and the broad outlines of projects that looked plausible for New York.

The meetings quickly tapped and then progressively generated enthusiasm among high city officials, increasing numbers of whom attended the annual Fit City conferences. One

DOHMHer, for example, remembered "fun meetings where we tossed around all sorts of ideas about stairs and windows and lines of sight, how sidewalks look, the ideas of a Danish architect who came to talk with us. It helped that DOHMH had no skin in the game: we could make the case, but we weren't anyone's competitor." Lee's framing and the AIA's engagement and convening paid off: agencies began to advance and accelerate strategies to promote active living within their own domains and in collaboration with others.

WALKING

For the city's energetic DOHMH, active living posed strategic challenges. One was organizational: the department's leaders well understood the health benefits of physical activity, but they had no direct responsibility for the supply side of the picture—the built environment, responsibility for the multiple parts of which spanned several departments: Design and Construction, Planning, Parks and Recreation, Buildings, Public Works, and above all, Transportation. Moreover, as noted above, the type of regulatory interventions they favored to promote healthy eating and discourage smoking—for example, taxes on cigarettes and requirements to post information on menus and food labels, which could be implemented through the city Health Code—could not easily be marshaled to shape the demand for active living. But the department's links with the AIA, which had hosted the above-mentioned series of annual Fit City meetings, paved a promising path for intervention and for partnerships with other city agencies.

A commitment made at a Fit City conference by David Burney, then commissioner of Design and Construction, inspired the agency partners to coalesce around the creation of "active design guidelines" that showed how buildings in the city could be made more amenable to walking. These guidelines, designed and issued in 2010 with the participation of DOHMH, Design and Construction, City Planning, DOT, and the AIA, focused on identifying evidence-based changes in the urban built environment that could promote physical activity through design—for example, improvements in the signage, lighting, and safety of stairwells in ways that would make the stairs a more attractive alternative to elevators and escalators. The premise—that the most promising way to change behavior is, as an official explained, to "make the healthy choice the easiest choice . . . to 'bake it in'"—exemplified what has come to be called nudging and has extended into the sphere of active living a resolve to make "the healthy choice the default social option" and to change "toxic environments," a mission that also animated the administration's approach to discouraging smoking and the consumption of unhealthy foods.[6]

These plans did not go unopposed—questions arose about security, liability in case of falls, and compliance with fire codes, for instance—and the city's Department of Buildings had to be won over. But, fortified by an executive order from Mayor Bloomberg (who encouragingly noted that he made a point of taking the stairs in his five-story home[7]) city officials began redesigning stairwells in their own buildings. They then hoped to incorporate the guidelines in new construction contracts for city buildings and eventually in local building codes—a broader and more contentious proposition that evoked concerns about cost and elicited (as an interviewee put it) "the resistance of the private sector to regulation of any sort." Meanwhile, the active living planners sought to raise the awareness and appeal of such (re)design projects by creating a

Center for Active Design, by expanding a program launched by the DOHMH team and AIA to train architects in the precepts and practices of active living, by securing foundation funds to support this instruction, by running training sessions for housing developers, by presenting and explaining the guidelines, by offering visual examples of them in pamphlets and other publications aimed at both professional and community groups and boards, and by seeking legislation to increase parity between stairs and elevators in the rules governing construction and renovation of buildings.

Another strategy, begun under Bloomberg and continued by de Blasio, sought to encourage walking by redesigning outdoor public spaces. As a step toward the European-style Complete Streets that the middle class was said increasingly to favor, the creation of car-free plazas in busy areas such as Times Square—"peoples' playgrounds" (as a proponent put it)—promised to promote both active living and social cohesion. As one might expect, a dramatic reconfiguration of space shared by pedestrians, motorists, and cyclists stirred controversy. Critics charged that traffic diverted from its customary channels would clog streets nearby and that businesses would suffer. Supporters rejoined that by increasing the volume of foot traffic, the plazas would encourage shoppers; that drivers would learn to adapt (or, better still, find ways other than cars to get around); and that the plazas would promote active living, social integration, and kindred urbane virtues.

In 2015 it appeared that proponents had the better of the argument—at least to a point. Business owners in and around Times Square, Herald Square, and other sites of pedestrian plazas generally welcomed the new layout. And the altered flow of traffic had not proved to be paralytic (at least no more so than Midtown's customary gridlock).

On the other hand, the social composition of the plaza at Times Square had integrated into the flow of residents and tourists some problematic characters—namely, people suited up as cartoon figures (for instance, Elmo of Sesame Street)—who badgered passersby to have a picture taken with them and then demanded exorbitant fees for the service. When Elmo and his confreres were joined by scantily clothed women ("desnudas") who entertained and hustled the crowd, the integrative slope grew too slippery, and critics stormed the pages of the city's tabloids to denounce the whole project and demand that the plazas return to the status quo ante. "Mister Mayor, tear down this mall," thundered Denis Hamill in the *Daily News*, a sentiment seconded by columnist Linda Stasi, who declared that "the ridiculous pedestrian malls" forced on the city by "former Mayor Bloombucks" had "turned Times Square into Tits Square."[8] Mayor de Blasio's police commissioner, William Bratton, admitted that he would prefer "to just dig the whole damn thing up and put it back the way it was." And de Blasio himself appeared to wobble, opining that the plazas had "pros and cons" and deserved a "fresh look" by a task force.[9]

Other voices accentuated the positive: since the plazas opened in 2009, city records showed that "pedestrian injuries dropped 35 percent . . . [and] injuries to drivers and passengers in cars fell 63 percent. . . . Business boomed. Surveys reported leaps in satisfaction by residents, workers and tourists."[10] Proponents of the plazas, including the heads of Transportation Alternatives and the Times Square Alliance (a group of local businesses) urged the city to stay the course and deal with unsavory interlopers by tighter regulation of the public spaces in which they congregated.[11] As these supportive voices grew more indignant and insistent, Mayor de Blasio distanced himself from his musings about a fresh look, and in 2016 city officials adopted regulations that created "designated activity zones" for commercial pursuits (such as those of Elmo and colleagues) and "pedestrian flow zones" for walkers.[12]

While sustaining the momentum of pedestrian-friendly initiatives launched by the Bloomberg team, de Blasio added another of his own, Vision Zero, a strategy he had endorsed in 2013 in the mayoral campaign, in which injuries and fatalities on the city's streets loomed large.[13] To enhance safety for pedestrians and cyclists, Vision Zero included a reduction of the legal speed limit within the five boroughs from thirty to twenty-five miles per hour, tougher legal penalties for drivers who injured or failed to yield to pedestrians and cyclists, and the redesign of especially dangerous streets and intersections.

The strategy, which won a $25 million grant from the federal government in 2014, reflected multiple and mixed motives. City hall relished the prospect of fewer lawsuits against the city. Families for Safe Streets, an organization akin to Mothers Against Drunk Driving, founded by relatives of pedestrians who had been struck down by cars, and described by an advocate as a "big game-changing interest group," fought hard to give the threat of "kills and serious injuries" high visibility. And Transportation Alternatives hoped that Vision Zero would bolster the case for a comprehensive strategy to develop Complete Streets—that is, streets that honored "the needs of all road users, including people walking, biking, taking public transportation and using automobiles, and people of all ages and abilities."[14] As usual, marketing of the initiative drew strength from causes and movements distinct from, but related to active living per se. "You get a lot farther selling safety than you do health," a participant remarked.

For two decades this selling had gone well in the city and beyond it. Pedestrian deaths nationwide between 1990 and 2009 had declined from 6,482 to 4,109. From 2009 to 2017, however, they rose by 45 percent, prompting a traffic expert to lament "a complete reversal of the progress that had been made."[15] In New York City, by happy contrast, pedestrian deaths continued to decline—from 144 in 2016 (slightly above the 139 deaths recorded in 2015) to 101 in 2017—the lowest number since 1910, when the city first recorded such events.[16] New York was certainly not immune to the factors most often cited to explain the national reversal: more SUVs and light trucks (more injurious in a crash); more trucks delivering packages ordered online; more congestion as Uber and other ride-sharing vehicles proliferated; and more drivers and pedestrians who are distracted by calls, texts, and news on their cell phones. Apparently Vision Zero was working, at least for walkers.

Moreover, even as the city seemed to dodge the threats that innovations in technology and transportation posed to the safety of pedestrians, two factors, also exogenous to active living policies, hinted at further progress. First, in February 2019 Mayor de Blasio and New York State's governor, Andrew Cuomo, agreed to implement congestion pricing on overabundant vehicles (such as taxis and ride-share cars) in designated parts of the city. The new pricing scheme—inspired by a congestion zone in London, which had since 2003 accelerated the speed of traffic, improved the quality of the air, and triggered increases in walking, biking, and the use of mass transit—conjured up, in the words of New York's DOT commissioner, Polly Trottenberg, "a lot of pretty transformative possibilities," including expanded sidewalks and pedestrian areas.[17] And second, a package of state and city laws, adopted in July 2019, that aimed to curtail greenhouse gas emissions reinforced the appeals of the active living agenda.[18]

BIKING

Defining the proper physical and social place of bicycles in a city with an urban grid as dense and a flow of traffic as relentless as are those in New York has long divided public opinion. Naysayers (such as the curmudgeon who, in a letter to the *Daily News* urged Mayor Bloomberg to "make it illegal to ride a bicycle in the city"[19]) would consign bicycles to quiet country lanes and discourage their presence on packed city streets in which cars constantly jockey for advantage by seizing any available morsel of roadway. Advocates replied that the city's structural lack of hospitality to bikes is precisely why its leaders should be open to change: making bicycles and bike lanes more accessible would discourage the use of cars and thereby reduce pollution of the air, the drain on fossil fuels, the number of automobile collisions, the incidence of deaths and injuries to pedestrians, and—neither most nor least salient on the list of benefits—would also lead to improved fitness and health among the (presumably growing number of) citizens who cycled.

A master plan adopted in 1997 envisioned the construction of nine hundred miles of bike lanes and paths in the five boroughs, most of them under the authority of the DOT. Advocates for alternative transportation, however, viewed the Giuliani administration as hostile to them, an attitude they ascribed as well to Iris Weinshall, Giuliani's commissioner of DOT, whom Bloomberg initially retained. Bike enthusiasts invidiously compared Bloomberg's DOHMH, within which percolated ambitious designs to combat smoking, trans fats, sugary drinks, and obesity, with his DOT, in which the expansion of bike lanes seemed far down the list of concerns.

In 2006 the bikers' frustrations spilled into the public domain when Andrew Vesselinovitch, DOT's bicycle program director, resigned and set forth in an op-ed piece in the *New York Times* his reasons for doing so.[20] The pace of installation of new lanes, of which 250 miles had been added between 1997 and 2004, had slowed to fewer than 20 miles between 2004 and 2006, he contended, adding that he found it unfortunate in a city in which half the population was overweight, "in part because of too little physical activity," and in which cars mightily depress the quality of the air that the DOT was "not truly committed to promoting bicycling."

A year later Weinshall stepped down as DOT commissioner and was replaced by Janette Sadik-Khan, a lawyer and former federal transportation executive, an ardent proponent of a bigger role for cyclists in local systems of transportation, and a plausible ally for active-living-minded officials in the DOHMH.[21] "Very different from the traditional traffic engineer," Sadik-Khan, said an admiring advocate, cleverly "wrangled the DOT bureaucracy" by proceeding on the assumption that "traffic engineers are smart, but they've been given the wrong problem to solve." Moreover, the new commissioner's wrangling took advantage of an unforeseen development that had triggered closer cooperation between the two agencies before she took office. In response to a demand by Transportation Alternatives that city hall study and report publicly on cyclist fatalities in the city, an official in DOHMH launched research on the issue, working "hand in hand" with DOT and with representatives of the police and the parks. "The process," a participant recalled, "was very transformative. We were four agencies plus Transportation

Alternatives. We analyzed those deaths in detail and wrote a report, which was a huge success and led to four commissioners standing in Central Park to announce a large expansion of bike lanes in New York City."

With Bloomberg's backing—and that of the public, 66 percent of which thought bike lanes a good idea[22]—Sadik-Khan announced plans to accelerate the installation of bike lanes in the five boroughs, a proposal that, as also was the case with the pedestrian plazas, elicited less than universal acclaim. Critics advanced the usual arguments: parking spaces, already scarce, would be fewer; the access of shoppers and workers to stores and offices would be encumbered; commutes would be slower and roads more congested; confrontations and collisions between cars and bikes would increase. Negotiations convened by city officials—including those in DOHMH district offices that were, an official explained, "designed to build trust"—with city councilors and neighborhood organizations tried to address these concerns and develop consent among residents affected. When plans for new bike lanes in a neighborhood in Brooklyn, for example, got a negative response, an official recalled how that district office "worked with community groups and the DOT. The office negotiated for the community, including its demands for improvements for pedestrians. As a result, they got more miles of bike lanes than they had expected." This cultivation of the grass roots was rewarded, and not only in Brooklyn. By 2015 the city had installed 1,000 miles of bike lanes, which it hoped to expand to 1,800 miles by 2030. Ridership, it noted, was up, and safety had improved.[23]

Between 2006 and 2019, bike trips in the city expanded from a daily average of 180,000 to 460,000—good news in that growing ridership brought cyclists both a stronger sense of safety in numbers and a larger constituency for cycle-friendly policies. On the other hand, the increase clashed with the same coincident trends that also imperiled pedestrians—more motor vehicles crowded the streets as the economy recovered, deliveries of goods ordered on line soared, and mopeds multiplied.[24] The proliferation of ride-sharing services much accentuated the congestion: consultant Bruce Schaller found that "for every mile of personal-car driving the [ride-sharing] companies remove from the road in large U.S. cities, they add 2.5 miles of driving in a ride-sharing vehicle." Indeed, surveys showed that "roughly 60% of riders in Ubers and Lyfts would have walked, biked, taken public transit or stayed home if a ride-hail car hadn't been available."[25] In the summer of 2019 the delicate equilibrium between cars and cyclists appeared to be collapsing: in July three cyclists were killed by crashes, bringing the number of such fatalities in 2019 to fifteen, more than had transpired in all of 2018 (ten).[26]

Unsurprisingly, headlines screamed ("Linking Death to Bike Battle") and advocates wondered "What Happened to the Bike Safety Push"?[27] The de Blasio administration deplored the carnage, assigned about one hundred new city transportation workers to address the issue, and pledged to speed the construction of new bike lanes (which occupied 1,243 miles amid the city's 6,000 miles of streets) from 20 to 30 miles each year.[28] Cycling advocates greeted these promises with restrained applause. More lanes were needed, to be sure, but the lanes should be *protected* from the flow of traffic (which was the case for only about one-third of the city's lanes in 2019). And the protection must be designed to avert accidents. If the lane "runs between a curb on the right and parked cars on the left . . . cars traveling to the left of the parked vehicles have no clear view of what may be coming along in the bike lane, and cyclists in that lane have limited ability to see a driver who turns right and fails to slow down enough to avoid a collision."[29] Nor is that all: lanes should be not only (properly) protected but also *connected*, thus relieving bikers of annoying stops and starts. The circle of political life whirled on: Each advance triggered more challenges, more demands, more promises, and more critical scrutiny of the city's willingness and capacity to meet them. "The devil," averred one advocate "is in the implementation."[30]

The city also made bikes more accessible to residents and tourists by means of a public–private partnership, the Citi Bike program, launched in 2013 with start-up funds from Citibank, under the oversight of DOT and the management of NYC Bike Share, a not-for-profit group. The program—said to be the city's "first new wide-scale public transportation option in more than half a century"[31]—let riders use a credit card to rent a bike at various sites around the city for a specified time and then return the bike to any Citi Bike location. Convinced that (in the words of former transportation commissioner Janette Sadik-Khan) "You can never do enough outreach," DOT sought to mobilize public consent (in 2012, 74 percent of respondents told pollsters that they supported Bike Share) by launching a "long term public dialogue" and "an exhaustive and highly participative planning process" that included 159 public meetings, presentations, and demonstrations; another 230 meetings with elected officials, property owners, and other stakeholders; an interactive station planning map that elicited more than ten thousand suggestions for siting; and the unveiling to the public, community boards, and other audiences of 2,881 "technically viable options for the 600 bike share stations."[32]

This assiduous outreach helped to inoculate the plan both against predictions of doom (comedian and *Daily Show* host Jon Stewart, for example, had sarcastically sketched a new business venture, his "Street Brain Material Removal Service"[33]) and against several practical complications that arose in the course of implementation—deciding how much to charge for rides and memberships, maintaining the bikes in good repair, liability for accidents, whether to require that riders wear helmets, and. not least important, how much direction the city government should assert over the program's private management. In the glare of a harsh critique published by the city comptroller's office in 2014, and the bankruptcy of the Montreal-based company that supplied the program's bicycles and other equipment, NYC Bike Share was taken over by a consortium of investors, given a new infusion by cash by Citibank and Morgan Stanley, and put under the control of Motivate, "a unique company focused solely on operating large-scale bike share programs."[34]

By late summer 2014 Citi Bike had not solved all the problems that dogged its early days—for instance, increases in the cost of a ride and of an annual membership had slowed the growth of ridership—but it was able to report that activity since the program began had surpassed 20 million miles. In July 2015 its leaders announced plans to double the number of bikes available (from six thousand to twelve thousand) by 2017 and to add new stations in Brooklyn, Queens, and upper Manhattan; the expansion of sites, stations, bikes, and riders continued steadily thereafter.[35] By the fall of 2019 the program had become by any reasonable standard "institutionalized": riders were using 14,000 bikes at 739 active stations in Manhattan, Brooklyn, Queens, and Jersey City for 2.5 million trips (on average 80,475 per day).[36]

PARKS AND RECREATION

New York's concrete jungle is studded with parks and green spaces, large and small, that offer opportunities for walking, biking, sports, and after-school programs. Everyone loves parks (which occupy about 14 percent of the city's land area), and "lower crime, improving park conditions, and a growing city" have encouraged ever-heavier use of them.[37] In the early 1980s Central Park welcomed about 12 million visitors but suffered roughly one thousand crimes per year. In 2017 visits had risen to 42 million and crimes had fallen to fewer than one hundred per year.[38]

The universal affection for parks, however, turns them into big green Rorschach tests that invite conflict over their character and contents. Where should they be located? Should policy emphasize the construction of new parks or the repair of older ones? What should be the relative priority and placement of, say, walkways, playgrounds, lawns, pools, pavilions, gardens, rest rooms, and benches (the latter a mixed blessing for active living aficionados given that, according to one survey, 61 percent of U.S. park-goers call themselves "sedentary").[39] And how much does size matter? Although very small parks need work "desperately," concentrating on them, said Tupper Thomas, executive director of New Yorkers for Parks, is "not a good idea." To make an impact on a community, the focus should be on "midsized parks. Having a beautiful park nearby, a place where you can meet up with friends or go for a run, is essential."[40]

The city's more prominent parks, moreover, are battlegrounds on which proponents of pedestrians and cyclists ally in the fight to limit cars. Although Transportation Alternatives has long argued for a complete ban on cars in the parks, its executive director praised as a "great leap forward" the announcement by Mayor de Blasio in 2015 that major portions of Central Park and Prospect Park would henceforth be off limits to cars on weekdays as well as weekends, thus creating "safe zones for kids to play in, for bikers, for joggers," and for others who want to "enjoy the park in peace."[41] Peace cannot be perpetual, however: parks are also sites of skirmishes between cyclists and pedestrians over how car-free space should be apportioned between them.

The complexities inherent in making policy for parks are aggravated by governance arrangements little less diffuse than are the expectations parks evoke. To make policy—and raise money—the Department of Parks and Recreation (DPR) works with various nonprofit conservancies, advocacy groups, private donors, partnerships, funds, trusts, "Friends of," and alliances; with local public officials (notably borough presidents and members of the city council, who can allot discretionary funds to parks in their jurisdictions); with community groups; and with consultants, architectural and design firms, and researchers.

Public agencies other than DPR also get into the act. Parks are important to the greenness that the Department of Environmental Protection seeks to promote, may figure in the design and maintenance of housing under the control of the New York City Housing Authority, are loci of instruction and recreation for school children the Department of Education (DOE) oversees, and are—of course—"at the center of [the] active, healthy living" the DOHMH encourages.[42]

Under Mayor Bloomberg the city invested heavily in parks and park-like settings. Commissioner Adrian Benepe installed "parks in schoolyards,. . . on old industrial 'brownfields' [and] . . . on derelict railroad spurs. Just about anyplace they could find to put in a patch of green space, they did."[43] Those endeavors more than doubled the capital spent on parks over the prior ten years, according to one expert's estimate[44]—a commitment dramatically exemplified by the High Line, a park of eight acres that runs for thirty-one blocks along the west side of Manhattan. Bloomberg's DPR also collaborated with DOHMH and DOE on the interagency planning that galvanized city agencies in support of active design guidelines and on after-school programs that incorporated physical activity.

The report *Take Care New York 2012: A Policy for a Healthier New York City*, which DOHMH issued in September 2009, promised cooperation among DOHMH, DOT, and DPR to ensure that all New Yorkers had access to "safe places near their homes where they can be physically active," including schools and parks.[45] The task force Bloomberg created to craft policies to reduce obesity recommended additions to the number of playground attendants who led Kids in Motion, a program that sponsors physical activity in playgrounds around the city.[46]

Equity claims among boroughs, neighborhoods, and income groups complicate parks policies, however. Communities that do not clamor for bike lanes may nonetheless view parkland as a welcome addition to their often limited settings for recreation and leisure, and their leaders sometimes grumble at the obstacles to acquiring such sites. In 2013, for example, city council member Fernando Cabrera of the Bronx complained not only that his borough had been ignored but also that the "participatory design process" and the city's rigid procurement rules had caused the DPR to lag in implementing projects for which he had earmarked discretionary funds.[47]

Convinced both that the active living initiatives of the Bloomberg team had centered too much on better off communities, especially in Manhattan, and that the city's poor have less access to venues for physical activity, including safe, clean, well-maintained public parks, Bill de Blasio, who succeeded Bloomberg as mayor in 2013, unveiled a generously funded Community Parks Initiative (CPI) as part of his "equity agenda."

To rewrite what de Blasio called a "tale of two cities," illustrated by Bloomberg's alleged preoccupation with urban designs that enticed tax-paying businesses and residents into the city, and especially into Manhattan, he pledged to bring a fairer share of the good things of municipal life to lower-income, less advantaged communities in all five boroughs. Noting that the city's spending of $5.7 billion on parks over two decades had included less than $250,000 for 215 of them, de Blasio earmarked within his budget ($1.27 billion for parks in fiscal year 2015) an allocation of $130 million over four years for the rebuilding of 35 heretofore neglected parks and for targeted improvements (mainly repairs and repainting) in others.[48] The selection process would favor parks that not only had received little capital investment in past years but also were situated in communities with exceptional rates of poverty, high density, rapid growth, and strong "local partner" organizations.[49]

This Community Parks Initiative (subtitled "A Framework for an Equitable Future"), drew praise from a wide range of local leaders for a long list of reasons, among which new opportunities for active living in disadvantaged communities were prominent though (as usual) not dominant. Public advocate Letitia James praised the initiative as "instrumental in the fight against childhood obesity" and other health issues prevalent in low-income communities. "No matter their zip code," declared Schools Chancellor Carmen Fariña, children "must have access to great parks, where they can exercise their bodies and minds." Assembly member Ron Kim praised the many benefits of improvements in the parks, "from fighting blight to facilitating physical activity and better health in our neighborhoods."[50]

A year later (in October 2015), de Blasio doubled the budget for the CPI to $285 million through 2019, an increase that would expand to sixty-seven the number of parks slated for reconstruction, enlarge the scope of targeted improvements to be made in other parks, and increase the ranks of personnel to maintain parks and conduct recreation programs and fitness classes in them.[51]

Because everyone loves parks, allocation of funds to build, rebuild, and improve them triggers classic distributive strategies (i.e., log rolling) that invite credit claiming by political leaders, including ones in city hall, community boards, and advocacy organizations. And the transition from macro (park policy) to micro (the design of individual parks) is no less fraught with tradeoffs among dispersed preferences. The community consultations that the de Blasio team valued highly and advertised widely disclosed sharp differences over what the CPI money should build, buy, and fix. In Lyons Square Playground in the South Bronx, for example, desiderata included bocce, a skate park, more basketball courts, grass and flowers, a dog run, a stage, and more. As the consultant guiding the plan worked to reconcile preferences with practicality, fitness-enhancing projects were sometimes sacrificed to other priorities, sometimes emerged front and center, and sometimes got embedded within other design features. The plan for the new, improved park would—if approved by the city's Public Design Commission and then by DPR—move the basketball courts (to reduce noise that bothered residents in nearby housing), add fitness equipment for adults, install a durable Ping-Pong table, relocate playground equipment (old and new) away from asthma-inducing traffic fumes, put in a new three-hundred-foot garden walk, and add a rest room.[52]

Occasionally communities and the planners who advise them opt to make the health benefits of active living the centerpiece of their plans for the design of parks, as illustrated by the Haven project, also in the South Bronx. In this case, the creation of green spaces, recreational paths, and streetscapes along an area of 1.3 miles was portrayed and sold to funders mainly as a route to better health. The project manager of the New York Restoration Project "kind of joke[d] that hospital systems will prescribe access to the parks," and the head of HealthxDesign, a firm that collaborated on the project, declared that her goal was "to show that the health of the area's residents can change because of design decisions."[53]

SCHOOLS

Schools, a built environment in which children spend considerable time, are of obvious strategic interest for advocates of active living. Educators can increase physical activity by teaching students about its benefits, by conducting classes that involve physical exertion, and by partnering with other institutions on projects that encourage students to exercise.

School leaders and staff may resist invitations and pressure to expand this programmatic terrain, however. Their core mission is to develop their charges' cognitive skills; grades, scores on standard exams, promotions to higher levels, and rates of admission to colleges measure their success in achieving that outcome. External demands—new measurement fads, targets set by the federal No Child Left Behind Act, and competition from charter schools, for example—increase the stress. Like health care, education is belabored by critiques that deplore "monopolies" run by "providers," and (like health care) the field is besieged by reformers itching to subject educators to market forces.[54]

Many school systems, however, enjoy considerable insulation from these slings and arrows and retain substantial discretion in tackling their tasks and allocating their time and money. Although they are formally accountable to state (and the federal) departments of education and informally to parents (individually and represented by organizations such as Parent Teachers Associations), in many communities (including Albuquerque and Sacramento) school boards, superintendents, principals, and teachers (who are sometimes tenured or unionized, or both) have been put beyond the direct control of general-purpose government and therefore occupy an institutional world separate from that of city hall. Needless to say, this structural gulf complicates plans to introduce changes from outside (or, indeed, from within) school systems.

New York City is by no means devoid of education professionals who yearn to practice their own brand of politics in benign isolation from policy currents in and around city hall, but their ability to do so diminished dramatically in 2002 when the state government acceded to one of the two top legislative priorities of newly elected Michael Bloomberg and put the city's schools under mayoral authority. (Bloomberg's other top priority was a ban on smoking.[55]) School leaders, willy-nilly—in a local governmental regime that highly valued both health-promoting interventions and interagency cooperation, and in a civic culture replete with nonprofit and private actors who were determined to influence education policy—took a fresh look at the place of physical education in their curricula.

To be sure these leaders did not work on a tabula rasa. Education Law 803 of the state of New York, interpreted in extensive rules issued by the Board of Regents and the state commissioner of education, specifies how much physical education schools must supply to their students (at least "120 minutes per calendar week exclusive of any time that may be required for dressing and showering"), how many credits for physical education students must accrue in order to graduate, and whether recess counts toward the requirements ("No").[56] And although the federal government prefers to encourage physical activity in schools by means of grants, not requirements, the Child Nutrition and WIC Reauthorization Act of 2004 instructs school districts with federally funded school lunch programs to develop policies to promote healthy eating and physical activity. To manage these state and federal requirements and the multiplying demands of a range of local actors, the DOE created an Office of School Wellness Programs, which oversees a sizable portfolio of school-based fitness programs, including a "health-related fitness curriculum," a standardized tool with which to assess the physical fitness of students ("FITNESSGRAM"), and a special program of physical activities for middle schoolers.[57]

Moreover, the city is richly endowed with institutions—private, nonprofit, and public—eager to increase the prominence of physical activity in the daily lives of school children. For example, in 2014 Nike, a maker of running shoes headquartered in Oregon, gave the city's Fund for Public Schools a grant of $1 million, which the fund would make available to the DOE for augmented physical education. And two national nonprofit organizations, School Wellness Councils and Walking School Bus, offered technical assistance on how the city might expand its stock of fitness-enhancing projects.

Within the public sector, DOE got technical assistance from DOHMH for the Move-to-Improve program, which trains teachers of kindergarten through third grade in "strategies for integrating activity into daily classroom schedules."[58] The educators collaborated with DOHMH on the Bike to School and We're Walking Here initiatives and on the federally and state-funded Safe Routes to School program (which Transportation Alternatives had long energetically promoted). DOE joined with the DPR to run Shape Up, which offers free drop-in fitness classes ("aerobics, yoga, Pilates, Zumba, and much more") in all five boroughs.[59] And, should the School Construction Authority (which the state created independently of the DOE in 1988 to reduce the risk of corruption in expensive school building projects) want to embrace active design features in its work, the city's Department of Design and Construction has at the ready guidelines that endorse, among other things, gymnatoriums—a "flex space" that usually offers "retractable and bleacher seating that enables the space to be used for . . . athletic events or as a performance and assembly space" and is therefore an asset for schools that lack room for both a gym and an auditorium.[60]

Making these marginal undertakings more prominent in the mission and mindset of DOE, an agency facing continual and intensifying scrutiny of its performance on an ever-expanding range of criteria that have little or nothing to do with fitness per se, is an arduous organizational stretch, however. How successfully the formidable list of interventions contributed to the outcomes everyone envisioned—a satisfactory supply of well-taught, well-equipped classes and activities that improve the fitness—hence, the health—of New York's public-school students in grades K–12—remains in dispute.

One study reported impressive progress in eliminating the negative—namely, a reduction of 44 percent in injuries to children in New York City whose schools participated in the SRS program—whereas census tracts with schools without SRS showed no decline.[61] Accentuating the positive proved to be harder, however. Advocates eager to apply active design guidelines in the schools sometimes met delays at the hands of community boards, the School Construction Authority, district leaders, and parent associations.[62] And a report by the city comptroller's office in 2015 contended that nearly one-third (32 percent) of the city's public schools lacked a full-time certified gym teacher and that 28 percent had no indoor space for physical exercise.[63] The critique was quickly taken up by the Citizens Committee for Children and the Women's City Club of New York (two prominent voluntary voices among the city's densely populated chorus of reformers and watchdogs), which seconded the comptroller's demand that the DOE "immediately conduct a system-wide assessment of physical education in these schools," which were disproportionally found in East Harlem, the South Bronx, and other disadvantaged areas.

The politics of active living policy in New York present a paradox: under mayors Bloomberg and de Blasio the city developed an assortment of active living programs that are impressive in their variety and scope (all the more so because noteworthy changes in the built environment were not eclipsed by the familiar strategic default option—namely, seeking to boost demand for activity within the status quo)—but at no time did active living as a distinct priority stand as a foremost priority on the agendas of any of the top political and institutional leaders under whose authority these initiatives took shape. Tom Farley's account of the aggressive, innovative interventions of the DOHMH under Bloomberg, for example, makes much of campaigns against smoking, trans fats, salt, and surgery drinks while promotion of physical activity gets barely a passing glance (the main references to the health benefits of exercise in Farley's book appear on pages 43, 134, 154, 185, 256, and 258). Nor did active living stand high on the agenda of Bloomberg's successor, Bill de Blasio, in whose state of the city address early in 2019 health issues (and, a fortiori, active living) took a back seat to workplace protections, housing, improvements in the subway system, and free early education.[64] The famous Kingdonian confluence of problem, policy, and political streams did not elevate active living above (so to speak) the middle rungs of the municipal agenda, so what subtler, subterranean forces propelled these muted priorities into policy? Unraveling the paradox requires attention to the politics by which enabling contexts created entrepreneurial niches for innovation.

In New York under the Bloomberg and de Blasio administrations several interlocking, mutually reinforcing contexts enabled and energized policies to promote active living. First, the city's prominence within the ranks of "world" and "global" cities induced within its leaders curiosity about the range of policies pursued both by peers such as London and Paris and by smaller but highly instructive cities such as Amsterdam and Copenhagen, which strive to honor New Urbanism, smart growth, sustainability, greenness, clean air, and—somewhere in the middle of the list—active living. Second, mayoral determination to make measurable improvements in the health of the city's population created space for active living among many other health-promoting strategies on the municipal agenda and kept alive the possibility that entrepreneurs within DOHMH might make manifest its latent appeals.

Third, a strong, media-savvy advocacy group, Transportation Alternatives, which represents both pedestrians and cyclists, elevated walking and biking amid the city's myriad priorities and found potent allies in organizations such as Families for Safe Streets. Fourth, the city's DOHMH had sufficient size and organizational sophistication to nourish a mix of activities aimed at promoting healthy behavior, including active living, which benefited from the entrepreneurial energies of an assistant commissioner, who in turn institutionalized active living initiatives by hiring a full-time specialist and then by building the city's first public-health oriented built-environment team to advance these initiatives.

Fifth, top-level atmospheric support (from the mayors' office on down) for health-promoting policies and projects encouraged managerial and interagency cooperation, which served to launch and secure those endeavors in and across a range of city agencies.[65] Strong high-level encouragement meant that failure to cooperate had palpable organizational consequences, and DOHMH, DOT, DPR, DDC, and DOE became, and have been encouraged to remain, collaborative sponsors of pedestrian-friendly building designs and plazas, traffic-calming measures, bike lanes, bike sharing, safe routes to school, and new opportunities for physical activity in the city's parks and schools. Against a priori expectation, these "priorities" moved from the margins of policy to institutionalization.

6

EVALUATION MEETS IMPLEMENTATION

The Struggle for the Real

DID IT WORK?

Captivating though they may be, accounts of the structures and processes that shaped efforts make the built environment more favorable to active living in five cities are surely mere preludes to the big question about the dependent variable: what *outcomes* did all this work produce? For many policymakers, policy analysts, evaluators, and foundation funders, "impact," "return on investment," "cost effectiveness," and "demonstrations" (preferably "scalable") are the proper proof of all possible policy puddings. The question is easy to pose but, alas, not at all easy to answer—or even, for that matter, to explicate.

First, what constitutes *working* is far from clear. The premise underpinning and uniting the highly varied interventions in the five cities studied here holds that a shift from more sedentary to more physically active lifestyles will, ceteris paribus, improve the health of those who change their ways. That general proposition has strong scientific evidence behind it—evidence arguably stronger than that enjoyed by any health promotion strategy other than cessation of smoking. But the practical corollary—that changing the built environment in ways that encourage physical activity in the course of daily life is a reliable road to better health outcomes—is rather more elusive than the premise might suggest. No wonder, then, that scientific evidence for the "place-based programs" that John MacDonald and colleagues prefer to demand-side approaches "remains in short supply in determining what works and is most effective for improving health and safety."[1]

"Scientific evidence" is of course always welcome, but because many confounding variables complicate the demonstration of links between specific policy initiatives to advance active living and the health status outcomes of individuals or of populations, the causal consequences (impact, returns, and so forth) of, say, adding bike lanes and widening sidewalks cannot be gauged with anything close to precision. As Cornelia Guell and her colleagues concluded in their study of collaboration between the health and transport sectors to encourage "active" commuting patterns in Cambridge, England:

> Evaluating the impact of complex interplays of social, environmental, and political-economic contexts on population health behavior produces complex evidence that can rarely be distilled into simple, definite answers. . . . Epidemiological evidence for the health benefits of physical activity had political traction but concrete evidence for the effects of particular interventions—that infrastructure A leads to health outcome B—was hard to come by.[2]

Causal disconnects are more or less intrinsic to evaluations of such projects, of which one can say little more than that they "worked" by creating new options for members of the community who chose to take advantage of them.

All built environments offer abundant "use"-related indicators to which data may be affixed—the number of bike lanes built; the number of riders on them; the number of downtown walkers on a given day and over time; the number of people using parks, trails, stairs; and on and on. With adequate baseline data, a researcher might estimate the benefits of these policies by measuring changes—increases, so one hopes—in rates of use (of lanes, of sidewalks, of parks) over time. Such data tend to be sparse, however, and even if they were abundant, judgments on how large an increase in use justifies the costs of the policies (which include time, labor, materials, and possible externalities visited on nonusers of the altered build environment) are necessarily subjective, as is the intensity of the preferences of users and nonusers. Time horizons further complicate such calculi. An expensively built but little-used bike lane (or network of lanes) hardly looks like a productive use of resources, but there may be something to the adage "if you build it, they will come."

One also should include in the calculus collective, even if largely symbolic, benefits such as the public endorsement of healthier living, reduced pollution, social integration, and other objectives of the movements for smart growth, New Urbanism, sustainability, green space, and (of course) active living. How does one quantify the value of such investments in the legitimacy of social change, which may take years to show results? And what is one to make of cases such as "Townville," a Danish city in which a commitment to "intersectoralism" (health in all policies) eluded implementation, leading instead to the "reproduction of abstract rhetoric and vague plans," but nonetheless served "as a way to display and support good intentions and hereby continue the process" as the multisectoral cast of characters groped their way toward policy.[3]

What, moreover, is the independent variable called *it*? In the case of active living, this category is nothing if not capacious. As the five cities reveal, the demand side of the promotion of active living includes mayoral pronouncements, festivals, and 5K runs; fairs; public relations campaigns conducted by health care organizations, insurers, and the media; wellness initiatives launched by employers; and much more. The supply side incorporates bike lanes, sidewalks, trails, opportunities to exercise in parks, Complete Streets, and Safe Routes to School, to name the more prominent encouragements to active living. Generalization rigorously connecting outcomes (improvements in health) to processes (rates of use) amid this hodge-podge of strategies is impossible because one cannot identify and parcel out distinct, discrete lines of influence. At best, one can try to gauge the benefits of one or another approach (or of a mix of them) by examining the particulars of what has proven to be workable in concrete local contexts. Doing so demands a style of "scientific evidence" tailored not to the dogmas of research manuals but rather to the concrete properties of the subject matter under investigation.

Jerome Bruner, noting that "validity . . . is an interpretative concept, not an exercise in research design," asks: "Are not interpretations preferable to causal explanations, particularly when the achievement of a causal explanation forces us to artificialize what we are studying to a point almost beyond recognition as representative of human life?"[4] Nearly three decades later, Jerry Muller echoed this theme:

> [Developing] valid metrics of success and failure requires a good deal of local knowledge, knowledge that may be of no use in other circumstances—to the chagrin of those who look for universal templates and formulae. The hard part is knowing *what* to count, and what the numbers you have counted actually *mean* in context. . . . To the extent that we try to develop performance metrics for *any* complex environment that is either unique or substantially different from other environments or

> organizations, standardized measures of performance will be inaccurate and deceptive.[5]

Finally, *did* (or *does*) compresses and foreshortens assessment of initiatives that may take years to launch—and then need additional years to take root, win constituents, and (perhaps) show results. Reflecting on the "developmental nature" of the implementation of federal efforts to nourish local compensatory education programs, Richard Elmore and Milbrey Wallin McLaughlin warn that "*it takes time* to establish a smoothly running, successful local project" and that premature assessments disserve "both policy and practice."[6] An assessment of program results ten years after the ALbD program was launched concluded that "the varying contexts, resources, and strategies across participating communities provide more questions than answers about the most effective approaches to fostering active living."[7] Because changes in the built environment are notoriously glacial, snapshot evaluations cannot do them justice. One can only look and listen, wait and watch—exactly what a bottom-line mind-set fixated on short-term impact and return-on-investment declines to do.

These caveats might seem to deny that sound policy analysis can ever anoint interventions as successful or indict them as worthless or worth too little to merit continued support—a fine formula for what David Stockman once called "the social pork barrel," indeed, for any poorly conceived product of the imaginations of policy entrepreneurs earnestly spending other people's money.[8] Opportunity costs are salient, to be sure, especially when economic growth lags and competition for more slowly growing public revenues intensifies, so tough-mindedness is no small public virtue, and it is one that depends heavily for its exercise on rigorous policy analysis.

This unimpeachable proposition itself depends heavily, however, on what one means by *rigorous.* After decades of uncritical equation of rigor with measurement, standardization, and quantification, critics now call more forcefully into question the merits of "methodolatry" and "metastasized quantification" as guides to the formulation of policy and underscore the advantages of "sense-making" evaluations that grapple closely with the meaning of policies in the contexts in which they are implemented.[9] So viewed, the "does it work?" question is best addressed not by a quest for parsimonious ("twenty-five words or less") answers but by up-close-and-personal investigation of the cultural, organizational, and political patterns in play in the settings in which policies unfold. Policy analysis, in short, might do well to dismount its high horse, "ascend," as Marx said, "to the particular," open itself to the "refreshments of surprise," and scrutinize the myriad contingencies and sensitivities that accompany the implementation of policy innovations.[10] This approach is, at any rate, indispensable to a fair assessment of policies to promote active living, which are exquisitely implementation-sensitive. And more generally: as scholars labor to reintroduce to the center stage of policy analysis "the state," "history" and other worthy intellectual foci, implementation would seem to deserve inclusion on the guest list waiting to be "brought back in."

ACTIVE LIVING AS IMPLEMENTATION-SENSITIVE POLICY

Policy determines politics. This proposition, which Theodore Lowi distilled from E. E. Schattsneider ("New policies create a new politics"), suggests that public policies have distinct political characters and that the outcomes of a policymaker's handiwork depend importantly on an insightful reading and interpretation of the political challenges to which policies give rise.[11] Analytic efforts to encompass the policy/politics nexus within a variety of general frameworks are of enduring conceptual interest but tangential to the theme that emerges from the case studies presented in previous chapters here.[12] That theme underscores—to take semantic inspiration from John Wennberg's studies of local variations in medical practices—the highly "implementation-sensitive" nature of policies that aim to promote active living by changing the built environment.[13]

Today, long after Woodrow Wilson's sharp distinction between politics and administration has largely lost favor among students of government, policy analysts sometimes persist in postulating a linear continuity between the making of policy and its implementation—a kind of regular order in which first things (setting the ends and means of policies) come first, whereupon implementers in public agencies and private organizations put (or are expected to put) the policymakers' designs faithfully into practice. Abundant studies suggest, however, that this orderly image may be more exception than rule; that the policy/implementation dichotomy often breaks down—indeed, disintegrates; and that much policy is of necessity made, so to speak, down the line and along the way. The pursuit of policies that support active living, especially when changes in the built environment are contemplated, would seem to be a poster child for the disorderly dynamics of what is here called implementation sensitivity.

Implementation-sensitive policies are vulnerable to sharp (and, in the nature of the case, largely unforeseeable) disconnections between the avowed aims of policies and the political complications that emerge as policymakers work to achieve those aims in practice. And it is precisely such disconnections, evident across the case chapters, that dissolve the policy paradoxes from which this study set out—namely, that policy objectives that are near-universally endorsed, appear on the agenda of most organizations engaged in health care policy, and face no major technical problems in envisioning the transition from ends to means should encounter so long a parade of obstacles in their path. The resolution of these paradoxes lies in the realms of culture (local particularism and the meaning of place), organization (intersectoral pluralism), and politics (political championship).

IMPLEMENTATION SENSITIVITY IN PRACTICE

The first element in the implementation sensitivity of active living policies is "local particularism," a force that has long inconvenienced and at times embarrassed policy analysts for three reasons. First, the particulars of local life perplex the measurement that is purportedly the precondition for sound management of policies. "Thick description" beclouds the simplifying aggregations that analysts (and reformers) think they need in order to characterize in "actionable" terms the problems at which they aim.

Second, localism complicates the crafting and application of general solutions. Insistence on accounting for "how things work (or may not work) around here" precludes any One Right Way, leaving only myriad ways that may or may not turn out to be "right" when they are implemented in distinct local circumstances. The ensuing ambiguity and uncertainty are dispiriting alike for policymakers who seek scalable solutions that can be succinctly conveyed and for policy analysts who hope to meet that demand.

Third, local particularism obscures the normative foundations of policy reform, especially in the United States, where localism can equate with diversity that in turn translates into disparities that are plausibly construed as inequities. Reformers have long sought national programs that aim for uniformity in the allocation of benefits—Social Security and Medicare, for example—largely because discretion left to state and local governments may enable or, indeed, encourage discrimination against groups with limited political and economic power.[14] But a society uneasy about the reach of big government does not easily accept policies that markedly curb or constrain the authority of levels that are purportedly closer to the people, so in large swaths of public policy the social and cultural particulars that shape the works and ways of local government are constant components of implementation sensitivity.

Built environments (and the activities that go on, or are expected to go on, within them) are spatial configurations attached to distinct cultural connotations (in a word, meanings) for those "concerned" with the spaces in question. The city is "a concatenation of man-made, willed things—things that add up to a texture of places."[15] Because the built environment is a social, no less than physical, construct, a priori uniformities of problem measurement, policy prescription, and political governance tend to fragment amid myriad contexts.

That such local connotations should complicate the implementation of active living initiatives may be prima facie puzzling, given that the cities examined here, like most cities in the United States, have long been caught up in general patterns of metropolitan development. After World War II, suburbanization gained ground as federal funding for the construction of new highways linking center cities and outlying areas and as easier financing of single-family homes induced better-off inner-city dwellers to migrate outward, leaving the inner cities in a "cycle of cumulative deterioration."[16] The decline in middle-class, tax-paying residents depleted urban coffers, leaving fewer resources for the services and amenities on which the cities' competitive appeals depended, which in turn gave remaining city residents more reasons to flee. This metropolitan pattern embodied important cultural emphases that remain very much alive today. "Sprawl" and the cars that sustain it are still widely understood to connote freedom, opportunity, choice, and mobility.

In the 1950s and 1960s city leaders sought federal funds for urban renewal projects that subsidized construction of new apartment buildings, offices, and commercial enterprises in central business districts and, infamously, disrupted many lower-income communities in the process. The early twenty-first century has seen fresh strategic variations on the retention theme as city leaders pitch their appeals to millennials—well-educated entrepreneurial types in their twenties and thirties who want to enjoy a cosmopolitan and (often) car-free life before they consider a move to the suburbs. In this context, active living has become a means to the end of economic development, and "walkable/bikeable" is widely viewed as a salient, perhaps indispensable item on the menu of inducements. In all five cities examined here, political and business leaders worried about their competitiveness in attracting millennials and elevated active living on their civic agendas in consequence. These general affirmations, however, soon ran aground on one of the paradoxes from which this study set out—to wit, that broad consensus on the merits of active living often crumbles under the strains of local implementation.

These strains derive from an unseemly conjunction of legal technicalities and community activism—a conjunction that comes inexorably with the policy territory that is active living. On the one hand, policies governing the built environment unfold within a massively yet intricately codified world. Proponents of, say, wider sidewalks or new bike lanes must situate their strategic yearnings within labyrinths of laws, regulations, zoning codes, subdivision ordinances, guidelines, criteria, and informal understandings that have accreted over many years through the interplay of political, bureaucratic, professional, and interest group influences and come under the superintendence and interpretive authority of departments of transportation, public works, and other agencies. Projects such as Safe Routes to School, moreover, typically proceed (or fail to proceed) within systems that are institutionally detached from city government. Even with a sizable corps of legal experts within a sympathetic public agency or a savvy organizational ally such as Transportation Alternatives in New York City, active living entrepreneurs are at no small disadvantage in summoning the staff, expertise, time, and energy to navigate these legal-technical mazes.

Moreover, proposals that surmount technical hurdles may be obstructed by displays of grassroots democracy as representatives (sometimes formally anointed, sometimes self-proclaimed) of neighborhood or block associations and community councils—or perhaps unattached but affronted local residents who raise their voices—object to the changes in question and attract media coverage and sympathetic responses of city council members or in city hall. Every inch of local space has its individual or collective owners and a range of meanings—spanning past memories and associations, present patterns of access, and future images of how the community should evolve—that these owners feel entitled to assert and defend. At times these images clash temporally and substantively within both individual citizens and the communities they compose, and the processes by which clarification and accommodation may (or may not) be reached can be heated and protracted. Nor are they easily short-circuited. "I live here" constitutes a weighty claim, founded on intense and legitimate preferences, that demands respect when residents rise, as in a Norman Rockwell tableau, to rebuff the usurpations of outsiders and elites who presume to know better.

Bike lanes installed without adequate consultation with residents on the streets affected may trigger anger, appeals to local officials, and threats to block similar projects unless the planners mend their ways (as in Louisville). The prospect of new bike lanes on busy streets causes anxiety about access of commuters and of emergency vehicles (as in Wilkes-Barre, for example) or may be opposed by residents seeking to defend their access to parks. (In New York a protest against new bike lanes near Brooklyn's Prospect Park was led by a former city transportation commissioner who is also the wife of U.S. senator Chuck Schumer.) Proposals to revise zoning codes in ways that better reflect urban "form" can antagonize residents who insist that the allegedly antiquated set of rules on the books faithfully reflects the venerable character of their community—or conversely, citizens who oppose new rules that curb the sprawl that purportedly honors the rugged individualism of the Old West (as in Albuquerque). In Sacramento plans to make a neighborhood more walkable and bikeable by denying approval for a car wash that would attract auto traffic were quashed by locals who insisted that the benefits (new jobs at the car wash) outweighed the costs (more traffic).

Such clashes arise not only from proposed modifications of the spatial status quo but also from the addition of new resources to it. What, for example, is the right mix of fields, games, gardens, benches, and walkways in a park scheduled for restoration, such as Lyons Square Park in the South Bronx? Are exercise stations a good idea or a waste of money along the trails around Wilkes-Barre? Should a new community center in Louisville be designed mainly for indoor or outdoor activities? Should funds for Safe Routes to School in Sacramento and Albuquerque be used to build infrastructure, hire new staff, or launch a marketing push to encourage walking and biking?

Moreover, public perceptions of safety pervade the willingness of residents to entertain and to use active-living enhancements.[17] The safety issue appeared prominently and repeatedly across the five sites examined here—for instance, in parental worries about the "stranger danger" that safe to routes to school allegedly invited in Sacramento; in trepidation about encountering drug users in downtown Wilkes-Barre; in doubts that one can safely navigate bicycles on Manhattan's congested streets; in beliefs that more walkable and bikeable streets in Louisville's lower-income neighborhoods would enhance the mobility of criminals; and in the unwillingness of residents of a close-in Albuquerque neighborhood to traverse a pedestrian tunnel that led to downtown.

These encounters between continuity and change are turf battles in the true sense of the term—battles over contested understandings of what the French call *l'âme du lieu*, or, as Harry Eyres put it, "the dear particularity of place."[18] Such contests both stress and strengthen the coherence of local cultures by catalyzing "interpretive procedures for adjudicating the different construals of reality that are inevitable in any diverse society."[19] The legal-technical side of the active living exercise demands skillful, persistent navigation of arcane texts. The communitarian side puts a premium on consultation, negotiation, and the painstaking building of trust. In short, implementing the active living "consensus" requires, in the words of James Agee, "a nose for [the] intensely specialized chemistry of local intuition."[20]

INTERSECTORAL PLURALISM (1): THE FEDERAL SYSTEM

A second reason why active living policies are deeply sensitive to the dynamics of implementation is that these policies galvanize a wide and diverse cast of stakeholders both within and outside the public sector.[21] Authority over these policies generally divides among several different levels of government; is often fragmented within each level among various agencies; and is influenced at each level by many private and nonprofit organizations. This institutional pluralism within and across sectors sometimes surprises health care actors and advocates, who, setting out from the proposition that active living policies are, or should be, mainly about promoting better health and, having found detours in the path of installing "health in all policies," may feel like anthropologists marooned on unfamiliar terrain among tribes who speak impenetrable languages and practice strange customs.[22]

The Federal Government

On the demand side of active living policy, federal leaders enjoy something of a bully pulpit (for instance, the oft-iterated advice of the surgeon general that people should walk briskly for thirty minutes several days each week), but on the supply side (the built environment) the feds are arguably more a problem than a solution. Federal legislation on transportation policy mainly channels to the fifty states funds with which to build, expand, and repair highways and bridges. These allocations—a textbook case of Lowi's "distributive" (pork-barreling) arena and Wilson's diffuse costs / concentrated benefits category—confer profits and jobs on the engineers, road builders, construction trades workers, and developers who constitute the highway lobby (the "sprawl community") and erect tangible bricks-and-mortar monuments to progress for which political leaders can claim credit in the eyes of their constituents. This theme—several decades of federal subsidies for policies that have left U.S. metropolitan areas car-centered and fragmented—runs strongly through a vast critical literature as well as the five case chapters presented here.

In this densely institutionalized context, the health benefits of walking and biking count for little, generating as they do scanty profits, jobs, and monuments to mass mobility. On the other hand, critics have for decades documented and dramatized both the social costs of these distributive patterns and the health gains that argue for active-living-friendly policies. Concomitantly, the mélange of movements such as smart growth, New Urbanism, sustainable development, environmental protection, greenness, and, to be sure, active living, has (as the case chapters illustrate) challenged the economic and cultural logic of sprawl in ways that increasingly register in the political calculi that drive federal transportation policy. Although the scales remain tipped toward the interests of motorists and of those who pave their way, the individual—and increasingly concerted—force of the recent movements has begun to generate countervailing power that is visible in the evolution of the central foci over decades of federal transportation policy—from roads to highways in the 1950s to these plus mass transit (urban and other) in the 1970s, to the "intermodal" inclusiveness that commenced in earnest with the Intermodal Surface Transportation Efficiency Act (ISTEA) in 1991 and has become a staple in transportation legislation since then.

The States

As recipients of federal largesse, state policymakers tend to reinforce the equation of transportation policy with roads and bridges and resist appeals for alternative transportation. But the state contexts in which active living advocates pursue their goals are far from homogenous. Louisville and Albuquerque are the largest cities in heavily rural states whose governors and legislators worry more about the loss of jobs and residents in sparsely populated areas than about the smartness of urban growth and the prevalence of good sidewalks and bike lanes. In Sacramento, by contrast, transportation as usual clashes with California's commitment to environmental protection and pollution abatement, a tension that casts a favorable light on the merits of alternative transportation. When, as in Wilkes-Barre, the agenda for changing the built environment avoids confrontation with highway interests by emphasizing the construction, extension, and connection of trails, a distinct bureaucratic authority comes into play: whereas sidewalks and bike lanes are marginal to the core mission of the state DOT, trails are well aligned with that of the Bureau of Recreation and Conservation within Pennsylvania's Department of Conservation and Natural Resources. New York City, too, is a special case because its formal power, fiscal capacity, and bureaucratic breadth and depth ensure its substantial independence of the state DOT and thus a smoother (anyway less cluttered) path for alternative transportation.

Within state DOTs, moreover, the policy picture is not static. Agency leaders and staff do indeed tend to defer to legislated purposes and regulatory readings that accompany federal transportation funds, to the norms of the engineering profession, and to the distributive expectations of federal and state elected officials. Traffic engineers no longer routinely dismiss pedestrians as "little bits of irritating sand gumming up the works of their smoothly humming traffic machines," however.[23] After the 1973 Federal Aid Highway Act served notice that the construction of roads was only one part of the mission of a DOT, the agencies began publishing studies on "beautification, environment, commuting, bus lanes, road design and urban transportation planning" and found it politic to hold public hearings and maintain dialogue with citizens about their plans.[24] In 1993 amendments to ISTEA expanded still further the range of constituencies and purposes to which the agencies felt obliged to pay homage, including vocal critics of "disruptive highway investment."[25]

By the dawn of the twenty-first century, state (and local) transportation departments often included at least one partisan of tougher planning and alternative transportation with whom proponents of active living policies outside DOT may bond and strategize. These allies work to legitimate their agenda by disseminating the growing body of academic studies that make a case (based on health and other desiderata) for alternative transportation and the manuals and guidelines, devised by the National Association of City Transportation Officials and kindred groups, that guide its design in practice. And they seek compatriots in city halls, city councils, county commissions, and regional planning bodies in hopes of shifting policies at the margins and of earmarking funds for small projects that may demonstrate the benefits of innovation.

Counties and Regions

In health policy, the roles of county governments and regional organizations usually take a back seat in discussions of intergovernmental relations, which tend to be dominated by the familiar trio of federal, state, and local levels. The built environment, however, shaped as it is by patterns of metropolitan development in which transportation networks loom large, often brings to the fore county and regional institutions, the forms and functions of which vary from place to place.

In New York the role of counties per se is negligible because state law makes each of the five boroughs its own county, so county policies are coterminous with those of the city as a whole. Since 2003 Louisville Metro government, defined by Kentucky as a "consolidated local government," has assumed most government functions for the whole of Jefferson County. The other three sites have the traditional city–county dyad—Wilkes-Barre / Luzerne County, Albuquerque / Bernalillo County, and Sacramento / Sacramento County—but the division of labor within and between city and county differs considerably.

The city of Wilkes-Barre contains departments of health, public works, parks, and planning and zoning but no department of transportation. Luzerne County has a transportation authority, a planning and zoning department, and a county branch (located in Wilkes-Barre) of the state's health centers but no department of public works or parks and recreation and no department of health. In the city of Albuquerque one finds departments of transit, parks and recreation, planning, and municipal development but no department of public works and an "office" of the state department of public health. Bernalillo County houses branches of the New Mexico departments of transportation and of public works plus a county office of senior and social services. Sacramento city deals with public works (including transportation), parks and recreation, and "community development," but lacks a department of health. Sacramento County has departments of health, transportation, and regional parks, but not public works planning and environmental review.

These city/county variations present proponents of active living with mixed constraints and opportunities that defy generalization. A priori, the vesting of important powers over the built environment in the hands of county institutions would seem to strengthen the influence of sprawling suburbs; in practice, it is not clear that making the case for active living in this context is harder than doing so against opposition from neighborhood defenders of local particularism within city limits. County bodies, moreover, may broaden the range of enabling contexts that offer entrepreneurial openings for active living advocates. One case in point (see chapter 1) is the Recreation Facilities Advisory Board of Luzerne County, to which two active living aficionados in Wilkes-Barre gained appointment and in which they have pushed for higher county funding for trails and parks.

Metropolitan planning organizations present a comparably mixed political picture. These federated entities, which include representatives of local governments and many other interests (including advocates for pedestrians and cyclists) within metropolitan areas in deliberations on transportation issues, have been required by federal law since 1962 and saw their roles expanded by ISTEA in 1991 and by subsequent federal legislation.

Advocates of active living sometimes view these metropolitan entities as another layer of bureaucracy to surmount and another institutional forum in which opponents of density, mixed land use, and the accommodation of pedestrians and cyclists within transportation networks can block change. Others, however, regard them as occasionally productive enabling contexts for entrepreneurial openings that favor the active living agenda.

The state of California, for instance, has vested in its councils of government (COGs) the power to write "blueprints" for the reduction of air pollution. The one adopted by the Sacramento Council of Government in 2004 strengthened the hand of supporters of active living by articulating "a smart growth vision for the region" that seeks to "integrate land use and transportation planning to curb sprawl, cut down on vehicle emissions and congestion . . . [and] give options for people to walk, bike, or take public transportation to work or play."[26] Although its role is advisory, Albuquerque's Mid-Region Council of Governments has become a magnet for active living proponents who seek to shape the views of elected officials in both city and county and thereby change the tenor of metropolitan development. And the active living enthusiasts on the board and staff of that COG have reached out to (and have been reached out to by) like-minded officials in the National Park Service, the state DOT, and a prominent local hospital.

The COGs in these sites offer congenial settings in which young staffers, often with degrees in public policy; a commitment to smart growth, sustainability, and other persuasions allied with active living; and concerns about the many costs of sprawl may conduct research (devising surveys, synthesizing studies); forecast trends and shifts in regional patterns of growth; generate diverse scenarios for coping with those patterns; apply, and help others to apply, for grants; draft analytical documents; participate in the crafting of plans; and observe and interact with important metropolitan stakeholders from diverse jurisdictions and perspectives on planning. The recommendations of these organizations, whose mission is to take a broad view of their remit across both space and time, are seldom dispositive, but they may be educative about what the preferences and practices of local particularists mean for the big metropolitan picture.

Cities

The major public-sector players in active living policies at the municipal level form a kind of hierarchy at the top of which sit departments of transportation and public works. In the middle tier are departments of planning and urban design, parks and recreation, and schools. Toward the bottom—much to the frustration of public health professionals—one finds departments of public health. But this general picture is, like almost everything about active living policy, subject to variation, the most striking example of which in this study is New York City, where the city's Department of Health and Mental Hygiene spearheaded the agenda of a mayor (Michael Bloomberg) who set as his top priority improvement in the health of the population (including, but by no means only, via active living interventions). This department, a large organization with many layers of expert staff (including legal counsel), in turn coordinated with the Department of Transportation, headed first by one, then by another, nationally prominent proponent of alternative transportation and active living. Moreover, and unique among this urban quintet, New York City's government had formal control of the school system.

In Wilkes-Barre, by extreme contrast, downtown business leaders and trail associations took the initiative in active living from a city department of health that suffered from severe cuts in budget and staff. How public health officials in Louisville's Department of Health and Albuquerque's (state) office of health approached active living depended on whether and how far incumbents of the commissioner's office viewed active living as a legitimate and salient "public health activity" and on the signals they received from city hall and from city and county councils. In Sacramento, the state government's commitment to better air quality and to the laws, regulations, blueprints, guidelines, and milestones that expressed it reinforced the promotion of active (nonmotorized) living on the agendas of the mayor and county and city health officials.

Even when public health leaders embraced active living as central to their mission and enjoyed positive feedback from political superiors, they could make little headway without cooperation from agencies—Departments of Transportation, Public Works, Planning, Parks and Recreation—that were, at least formally, in charge of the built environment. The alliance-building this cooperation presupposed seldom came easily. The partners had to become, if not fluent, at least competent to speak each other's programmatic languages; were often, if not isolated, at least marginal in their own respective agency milieux; were compelled to fight, if not to reverse, at least to revise, settled arrangements that shuttled funds down the federal system to build roads and bridges, not sidewalks, bike lanes, and parks; and depended on securing, if not the ardent enthusiasm, at least the guarded endorsement, of local elected officials not only for simple demand-increasing projects (for instance, mayor's 5K runs) but also for complex redesigns of the built environment. Giving institutional heft and durability to the mélange of movements that sustained active living was a formidable exercise, and such success as these cities achieved in doing so largely reflects the adroitness and persistence of leaders and followers in these intersectoral alliances and accommodations.

The entry points that city halls, city councils, and county and regional entities offered to youngish proponents of active living deserves special underscoring because millennials are important not only as residents whom cities seek to attract but also as forces for change within the policy process itself. A major component of effective intersectoral collaboration is the gradual accumulation of progressive young outsiders who come inside to work within supportive organizational environments in which they can promote, for example, alternative (i.e., active) transportation. The presence and, in some cases, growing concentration of young policy staffers at work on a range of issues within congenial institutional settings has energized active living initiatives in all the cities examined here.

INTERSECTORAL PLURALISM (2): THE PRIVATE AND NOT-FOR-PROFIT SECTORS

The behavior of the populous universe of public stakeholders in the active living arena is ever-permeable to the influence of a well-stocked set of interests (many well-organized, some less so) in the private and not-for-profit sectors of society. As appears repeatedly in the case chapters, the built environment is of major material importance to road builders, architects, engineers, construction firms, real estate developers, real estate agents, and others. Because, as noted above, these actors typically participate in long established, well-heeled, and expertly staffed peak associations at the federal and state as well as the local levels of government, their political clout registers both up and down the federal system and at each level within it. Innovations proposed by active living advocates often entail more extensive public regulation and otherwise threaten the profits of these stakeholders—or at least promise to disrupt the practices by which they acquire those profits—and therefore trigger pressure on executives, legislators, and bureaucrats to resist, delay, or soften them.

Although active living proponents sometimes tar this whole collection of interests with one sprawling brush, some entrepreneurs—especially developers and real estate agents—have come to view active living as a route to doing well by doing good—a way to capitalize on growing demand for residences (urban and suburban) that are close to green spaces and feature opportunities for walking and biking. The importance of these innovations cannot be gauged solely by the current proportion of such projects in the total developmental mix, for they offer models that may gain national attention when advertised and disseminated not only by enterprising real estate agents within local markets but also by a growing cadre of nonprofit organizations, friendly to active living and its allied movements and increasingly creative in the arts of visualization, that is, at publicizing to audiences, both general and professional, pictures of a brave new walkable, bikeable world—pictures that may indeed be worth a thousand words.[27]

The progress of active living policies is in many ways a story of the proliferation and expansion of nonprofit organizations that have encroached on policy turf formerly ceded to for-profit firms in the sprawl industry. Some of these groups have focused narrowly on one or another ingredient of active living within their communities—for example, Biking for Louisville and the trail associations in and around Wilkes-Barre. Some promote a portfolio of options—for instance, Transportation Alternatives in New York City, an omnibus advocate for walking, biking, and mass transit. And some embrace active living within broader missions. Cases in point are the widespread New Urbanist, sustainability, and smart growth groups (the alliance of Walk Sacramento and Design 4 Active Sacramento, for instance); Friends of Parks in New York, and Wilkes-Barre's energetic association of downtown businesses.

Whatever their organizational pedigree and scope, these groups search up and down the federal system for allies—often a single like-minded official—in city halls and city and county councils; among local, state, and federal elected officials; and in local, state, and federal agencies, in hopes of finding enabling contexts in which to incubate proposals and projects that, in time, a fortuitous entrepreneurial opening—a grant, a shift in electoral fortunes, a reorganization of government responsibilities, the job promotion of an internal ally, or a helpful wave of media coverage, for instance—may activate politically. Sometimes, too, these groups combine in coalitions although, given the diversity of their aims, memberships, and depth of commitment, divisions over who gets (and gets to do) what should their advocacy pay off may strain their cohesion.

In the cities studied here, nonprofit groups (sometimes abetted by freethinking allies among for-profit developers and real estate agents) are indeed exerting countervailing power on behalf of active living at (and occasionally beyond) the margins of policies on the built environment. In contrast with the lobbies these progressive forces challenge, however, active living groups labor under serious political limitations—namely, heterogeneous constituencies, multiple aims, a paucity of material resources with which to work, and late entry into a settled, durable, and high-stakes policy game.

POLITICAL CHAMPIONSHIP

The balance of power between the forces inhibiting and those favoring active living initiatives varies over time within and between communities. Precisely because the factors that shape (or misshape) these initiatives are several and diverse, outcomes depend on whether and how much supporters of active living can mobilize political champions to weigh in and tip the balance their way.

By far the most important political champions for active living in the cities examined here are mayors, who came to endorse and promote it for three reasons. The first is its contribution as an economic development strategy. The mayors of the first decade of the twenty-first century, like their counterparts in the 1950s, struggle to keep or make their cities (and especially their central cities and downtowns) competitive with suburbs as places in which to live, shop, visit, run a business—and (a more distinctly twenty-first century theme) serve as magnets for entrepreneurs, innovators, and members of the "creative class." Of professional necessity, so to speak, they therefore canvass long lists of civic enhancements, including opportunities for active living.[28]

This strategy easily meets all three of Kingdon's conditions for agenda-setting:[29] it addresses an acknowledged problem (how to make central cities more appealing to middle- and upper-class types with money to spend); proposes policies sketched at length and in detail by articulate voices in the several sectors discussed in the preceding section; and carries sufficient political appeal that indifference to it would make local leaders appear to be out of some important (albeit amorphous) strategic loop.

A great many items are "on the agenda," however, and no political law says they cannot simply sit there indefinitely. Mayors, like other political leaders, are obliged to strike a balance between the symbolic and substantive dimensions of their commitment. So a second reason why mayors champion active living is that well-advertised attention to its demand side seems, at least initially, to strike a serviceable (and low-cost) balance. City halls in all five sites radiated sage advice about the virtues of walking, biking, and taking the stairs; prominently put their imprimatur on mini-marathons, bike races, and treasure hunts along trails; reminded residents that the city's glorious natural resources (parks, loops, waterfront paths) offered abundant opportunities for exercise; and publicly affixed their title to sections of municipal terrain ("mayor's miles").

Successful agenda-setting and demand-promoting, however, by no means presage linear evolution from full-throated political support to tangible changes in policy on the supply side of active living. Such changes carry costs: using finite political capital to build the trust necessary to negotiate with and to resolve conflicts among the multiple and sometimes discordant constituencies that animate local particularism and intersectoral pluralism (pedestrians, cyclists, people who want playgrounds and gyms, and so forth); finding the time amid crowded civic agendas for planning and winning the consent of stakeholders for plans that may seem feasible; raising money from grants, or finding it within constrained public budgets; waiting patiently, perhaps for years, to see tangible results.

These inhibitions notwithstanding, a third factor was at work across the five cities, namely, cautious optimism that the benefit of policies to reshape the built environment—improvement in the political economies of the cities' core (and perhaps in the health of local residents)—would justify these costs. One motivational size did not fit all, however, and the variations—and contingencies—in how these mayors came to champion this strategy are striking.

In Louisville, two bike-friendly mayors in succession, closely advised by dedicated cyclists, heartily embraced bike lanes and related encouragements to cycling as a hallmark of the city's commitment to active transportation. Wilkes-Barre's mayor between 2004 and 2012 concurred with local business leaders that a more walkable downtown was key to making residents "believe" again in the city. Albuquerque saw the replacement of a density-averse mayor by one who came to view a denser, multiuse, walkable, and bikeable downtown as a tool for stemming the departure of younger residents to more "swingin'" towns. New York Mayor Michael Bloomberg, powerful, energetic, and determined to improve the health of the city's population not only over time but while he held office, recognized how active living initiatives complemented antismoking and healthy eating interventions. Leaders in both the county and city of Sacramento wanted the capital city of the state that has led the nation in developing alternative transportation in order to combat pollution to walk the proverbial walk. None of these local leaders was prepared to dig up highways, tear down bridges, or otherwise break decisively with the car-centric customs that continue to dominate metropolitan development. All of them, however, added increments of political weight to the active living side of the scale and thus by their championship augmented the slowly but steadily accumulating stock of countervailing power.

Political championship in these cities reflected mayoral calculations that a nod to active living made sense both on its merits and as a means to civic revitalization, but their support was not sui generis. All five cities improvised active living "regimes" that cut (mutatis mutandis) across agencies, sectors, and levels of the federal system.[30] Active living supporters in Louisville, for example, helped to open doors to the mayor's office for actors such as Biking for Louisville and the ALbD grantees. In Wilkes-Barre, which struggled to fund municipal staples such as police and fire protection and sanitation, business leaders labored to keep clean, well-lighted sidewalks on the mayor's to-do list.

Active living in Albuquerque benefited from adroit advocacy by a handful of members of the city council, one of whom, an architect and planner by profession, pressed the case for Great Streets for seven years before the proposal was finally enacted. Had New York's Mayor Bloomberg somehow forgotten the role of active living in improving the health of the city's population, Transportation Alternatives and active living aficionados in the Departments of Health and Transportation would have reminded him. WALKSacramento built links within the city and across policy sectors at the city, county, and regional levels—a quest for "walking (and biking) in all policies," as it were, that both backstopped and stimulated the active living designs of the region's elected officials. In all these cities, networked political exertions contributed heavily to the production of political championship at the top.

"DID IT WORK" REDUX

The review just presented of what the five sites did and how they did it may be taken as an extended commentary on (even if not an answer to) the "did it work?" question with which this chapter began. Acolytes of Jane Jacobs and activists combating an obesity "epidemic" may dismiss the painstaking political crafting of policies examined here as too little, too late, another sorry instance of the proclivity of policymakers to fiddle while Rome burns.

To be sure, the rising influence of active living has not engendered in the cities in this study an emerging (still less a pronounced) resemblance to Amsterdam and Copenhagen. And the cresting taste for active living in European cities—the mayor of Paris, for instance, envisions "a bike lane on every street by

2024," a ban on all combustion-engineered cars by 2030, and, ideally, a city in which home, work, and activities all sit within a fifteen-minute "radius" of each other—makes one wonder what more it would take for U.S. counterparts to enter this enlightened circle.[31]

Cities in the United States increasingly share with European peers a "qualitative shift in cultural attitudes and preferences," especially among the young, toward "less reliance on the automobile and increased demand for living in mixed-use, compact developments in or near the city centre."[32] This mind-set rejects the incursions of highways into center cities; favors dense, vibrant, and car-free urban lifestyles; and expects local leaders to cultivate options for walking, biking, and use of public transit. American cities also draw (albeit much more sparingly) on strategic repertories that encourage walking (car-free pedestrian zones, traffic calming, wide sidewalks with amenities, improved signage, and easier street crossing) and biking (bike route networks, on-road bike lanes, bike priority streets, way-finding signs, requirements that new buildings provide a minimum number of parking spaces for bikes, and bike-share systems, for example).[33]

Important contextual differences constrain such shifts in "transport culture" in the United States, however.[34] In Europe, high costs of fuel and for licensing and parking of cars discourage their use. And although the role of political championship of active living in Europe has been little studied, a recent look at Vienna documents a range of effective political strategies that Viennese public leaders have pursued for decades and that U.S. proponents of active living might seek to emulate. In a nutshell, "implementation of sustainable transport policies . . . has been a long-term, multi-stage process requiring compromises, political deals, trial and error, and coalition-building among political parties and groups of stakeholders."[35] Political strategies, pursued and refined by the city's political and party leaders over decades, include the launching of small pilot projects followed by the expansion of those that prove effective and popular; a careful quest for public approval by means of referenda, both city- and district-wide; the "democratizing" of policy decisions by means of town hall meetings and the engagement of district councils; the "implementation of a multi-modal package of policies providing excellent alternatives to the car"; the assurance that most local groups derive some sort of benefit from active living endeavors; and—not least important—the successful pursuit of "extraordinarily generous financial support and regulatory cooperation of the federal government."[36]

The cultural and political contexts that enable the progress of active living in Europe are, although not merely embryonic, substantially less evolved in American cities, the achievements of which cannot be judged in detachment from those contexts. And so, although solidarity for motorists did not run deep within the veins of active living advocates, none of the cities studied here became a battlefield in a war on cars, as some ardent critics of car culture may have hoped. Rather, these active living projects accented the expansion of *choice* among multiple modes of transportation, thus avoiding, at least in principle, zero-sum conflicts that would have served their cause poorly.

Like it or not, many millions of Americans (especially, but not only, in rural and small- or medium-sized communities) depend on cars for work, leisure, and other activities of daily life and resent attempts from on high to make motoring more costly and less convenient. Projects that enable and encourage a choice to walk and bike by those inclined to do so advance the active living project without aggravating a social division that Paul Collier identifies as central to the anxieties that afflict contemporary capitalism, namely, a "geographic divide" that pits "regions . . . against the metropolis"—a tension that plays out forcefully between cities and their suburbs and satellites and in urban–rural tensions within states.[37]

All five cities tackled both the demand and supply sides of the promotion of active living, each with a "political metabolism" shaped by forces of culture, organization, and politics.[38] The specific aims of this exercise differed across the sites, as did the outcomes achieved to date. In no case did the accomplishments come close to fulfilling all (or most, or, arguably, much) of the ideal agenda promulgated by advocates of active living. But in no case did these results fail to move the proverbial needle somewhat closer to that agenda. By trial and error, leaders in each city read the civic (and metropolitan) scene for enabling contexts, interpreted the meaning of those contexts for entrepreneurial openings, and improvised programs and projects strong enough to weather the exigencies of the political exercise that effective promotion of active living demands. In this sense—the only sense in which these efforts can be fairly and realistically evaluated—their work worked.

Those frustrated by the multifactorial character of it all and seeking relief in a parsimonious checklist of "predisposing" variables might shine their spotlight on New York City, which displayed several distinct advantages, namely: (1) a civic culture that gave substantial scope to alternative (including active) transportation—walking, public transit, and so on; (2) a mayor (Michael Bloomberg) who made improvements in health of the population his top priority during his twelve years in office (and a successor, Bill de Blasio, who remained supportive of that goal); (3) city agencies (especially, but not only, the Departments of Health and Transportation) with staff, resources, and self-confidence sufficient to launch and sustain active living initiatives; (4) a cadre of advocacy organizations (above all, Transportation Alternatives, which represented both pedestrians and cyclists) that pushed strongly for active living measures; (5) an abundance of parks and other green spaces, large and small, that can house, across all five boroughs, facilities and activities that encourage physical exercise; (6) a school system under mayoral control; (7) recognition of the need for, and adroitness in pursuing, outreach and negotiations with community leaders and "voices" in the five boroughs; and (8) considerable independence from state government as city leaders went about the promotion of active living. What worked for New York worked in and by virtue of these distinct sociopolitical circumstances, as was also the case for the other smaller sites, which proceeded, each in its fashion, in institutional settings that both resembled and (mostly) differed from those in New York. In Sacramento, for example, strong legislative and regulatory pressure by state government, far from encumbering active living initiatives, energized regional, county, and local actors who viewed them as important components of the quest for cleaner air.

Finally, for readers seeking a succinct account not only of considerations of structure and process but also of outcomes—the elevator speech capturing the essence of why society should continue to occupy itself with the promotion of active living—one can safely say that these five communities, each in its own way, chalked up three formidable accomplishments as they struggled with the sensitive and contingent implementation of promising proposals. First, they added to the standard stock of health-promoting strategies new interventions not only on the demand side (more frequent exhortations from more numerous authoritative sources calling for more physical exercise) but also on the supply side (a built environment with better sidewalks, more bike lanes, and the like), which, if used intelligently and persistently by more citizens, may lead to improvements in the health of individuals and local populations. Second, they responded to intensifying demands by a growing cadre of residents for more abundant and appealing options for exercise (including active transportation)—no small achievement in an age that venerates choice. And third, they exerted a measure of countervailing power against a hitherto dominant pattern of metropolitan development that carried (and continues to carry) heavy social costs. By challenging the sprawling proclivities (logistical and political) of the car culture, the active living persuasion may be demonstrating what Max Weber wrote about cultural contingencies in general—namely, that they serve to load the dice, thus ensuring, as Jefferson Cowie writes, not that "the dice will always turn up a certain way, but [that] . . . they are more likely to do so."[39]

CONCLUSION

Active Living and Beyond: Bringing Implementation Back Into Health Promotion

Implementation is what happens to policies once they leave the hands of their formulators and another set of policymakers/managers must decide what those policies mean, here and now, for this citizen or class of citizens, for this community or collection of communities. As an alternative to (or replacement for) older coinages such as "administration" or "execution," "implementation" suggests that the transition from policy to outcome may not be linear and clear-cut—that some, perhaps much, policy gets made "along the way," taking shape on a kind of conveyor belt as it enters and leaves the hands of diverse artisans.

An implementation-sensitive policy is one that, however carefully crafted and articulated at the point of origin, takes shape contingently (that is, subject to no small measure of happenstance) in bits and pieces in distinct areal and institutional contexts. Active living is such a policy because its designs cannot be detached from locally particular meanings about space and place, about what is mine and what is ours; because it falls within the remit of diverse organizations at distinct levels of the federal system and across the public, private, and for-profit sectors; and because the trust, consultation, negotiation, and risk-taking required to legitimate change in local and organizational preferences and priorities rarely come to fruition without championship by political leaders beyond those responsible for launching a legislative or regulatory initiative.

To be sure, one can conceive circumstances that could absorb much of the implementation sensitivity of active living initiatives into policies that are more centralized and standardized. If, for example, climate change, soaring gas prices, or a serious national political shift to the Left that triggers a Green New Deal emboldened the federal government to reprogram large sums of money from highway construction to the building of sidewalks and bike lanes, with detailed statutory and regulatory strings attached to the funds to ensure that they are spent well (as defined by active living authorities), presumably local particularists, organizational contrarians, and local political leaders would jump as high as required to secure the funds. And yet it is hard to imagine the U.S. federal government, however committed to "seeing like a state" and to subordinating subnational pastiches to a governing national vision of active living, immunizing the project from much of the implementation sensitivity found in the five cities studied here.[1]

Active living, however stubbornly allergic to central planning and governance, may of course be an outlier—a policy that occupies a sparsely populated point on the extreme end of a continuum of complexity and contingency—and may therefore be a case of merely idiosyncratic interest. Or maybe not. Social Security would seem to be largely implementation-insensitive. Making policy about who gets what pensions is complex to be sure, but implementing the policy is straightforward: "assessing eligibility, determining benefits, writing checks." Nonetheless, the program's founders took pains to recruit local managers imbued with a "client-serving ethic," one emphasizing not only technical efficiency but also a "commitment to providing service."[2] Moreover, determinations of need in the Supplemental Security Income program and of disability in Social Security Disability Insurance, which require decisions by physician gatekeepers as to whether a given medical condition meets the criteria of Social Security (and Medicare) for coverage, are highly implementation-sensitive, and adapting to their managerial demands created "near chaos" in the Social Security Administration.[3] Nor have the strains of implementation entirely abated. For example, the growth of the disability program has triggered insistent demands that program managers curb fraud and abuse, including an estimated "$3.4 billon in overpayments to disability insurance beneficiaries in 2017, in part because of their failure to report work activities."[4]

Medicare also wrestles with national and regional coverage decisions and cost-containing features such as pay-for-performance, accountable care organizations, and price setting for physicians, none of which unfold without friction.[5] The Medicaid program is implementation-sensitive by design: the eligibility and benefits of the program vary considerably across the fifty states, and who gets what depends heavily on decisions by officials in state, county, and local offices of social services, which are shaped in turn by political signals from governors and legislators.[6] These and other programs that are broadly "welfare state" in nature—including the earned income tax credit, the supplementary nutrition assistance program, and temporary assistance to needy families—have a deceptive "in/out," off/on" appearance that cloaks the considerable discretion that those charged with "simply administering the law" often enjoy.

Another sizable set of programs falls under the rubric of "street-level bureaucracy."[7] In this case the problem is widely understood to be the impossibility of simply administering laws that are inescapably subject to the judgment of professionals (police, educators, and physicians, for example) whose discretion is seldom amenable to traditional hierarchical monitoring and supervision. Such policies (the universe of which coincides roughly with William Baumol's "unprogressive" part of the economy, in which technology cannot easily be substituted for labor[8]) are in the nature of the case sensitive to—indeed, customized by—implementation. Even more so are policies such as those that unfold within "confidential bureaucracies" at the core of the "poverty industry"—those that run foster care and private prisons, for instance.[9]

Distinct in turn from welfare-state and street-level type programs are the central subject of this study, ones that seek to shape the built environment. This is the domain of engineering—experts draw detailed plans and implementers set them in concrete. As James C. Scott has shown with examples ranging from the reconfiguration of forests to the construction of utopian cities, efforts to impose detailed blueprints from on high often fail to realize the planners' vision.[10] The local objects of the exercise tend to have both ideas of their own and powers (often subtle and unanticipated) to bring carefully laid plans to a halt or markedly to reshape them. Such disconnects also apply to the more modest endeavors, explored here, to make cities conducive to active living. In short, even a cursory canvas of diverse program types suffices to show the more than occasional salience of implementation sensitivity.

If implementation sensitivity is a near-chronic condition, it might seem surprising that analysts who focus on health policy, public policy, and the politics of policy have so little that is useful to say about it. Indeed, the analytical terrain suffers from (in Kenneth Burke's coinage) "exceptional existential poverty."[11] The explanation is that the three dominant factors in the case of active living—local particularism, intersectoral pluralism, and political championship—rest on considerations of, respectively, culture, organization, and politics, but these domains hold little conceptual resonance or traction for the epidemiologists, clinicians, and economists who rule the roost in health policy analysis and health services research, or for the economists and applied public policy analysts who study and teach public policy. (A rare exception from a luminous source is the recommendation of Esther Duflo that her fellow economists learn to think like plumbers who explore the inner workings of policies.[12])

Management experts in health and public policy analysis make much of organizational culture but little of culture beyond the walls of the organizations in question, and their approach to organizational behavior tends to favor business school nostrums that draw sparingly from the rich social science literature on organizations other than business firms. These approaches, aiming to market policy analysis as both scientific (adorned with hypotheses, "tests," large n's, regressions, and replication) and accessible (conveyable to policymakers in a short list of bullet points), promise both rigor and relevance and deliver neither.

In principle one might expect more of political science. Political scientists (anyway, analysts of the politics of policy), having awakened to implementation as a distinct policy process in the 1960s, watched it become, as Peter deLeon and Linda deLeon contended, "perhaps the policy analysis growth industry over the last thirty years."[13] Research piled up across three scholarly "generations" (focused, respectively, on implementation failures; on the need to introduce a bottom-up, community based orientation; and on the quest for a more scientific and theoretical approach).[14] But in 2002 the deLeons lamented that, as a field, implementation seemed to have "come and gone like an elusive spirit" and appeared to some to be on its "intellectual deathbed."[15]

A decade later, resurrection had stalled: indeed it had become "en vogue . . . to consider implementation as an obsolete research theme."[16] Irony abounds: One reason why "progress since 1990 has been much slower" is the "overly *ambitious* and *demanding*" nature of the theory-minded third generation research paradigm.[17] Another factor is the rise of "implementation science," crowned with an eponymous journal, which focuses on the "replication of randomized control trial results," relies heavily on "quantitative methods in the health sciences," and thereby aggravates the "dissipation" of implementation studies within the "core discipline of public affairs."[18]

The pretensions to implementation science have revealed little more than that "one size never fits all, that context matters, and that when we face an extremely complex condition, we are better off if we try to understand the particular issues than if we propose some form of generic metatheory."[19] The practical implications were, or ought to have been, humbling. Models that purported to capture the general features of the process were too abstract to yield more than restatements of the many ways in which implementation is problematic and challenging, prompting Richard Matland to bemoan the "lack of parsimony" dogging the development of "theoretical structure," and Harald Saetren to lament that despite the "exponential growth" of implementation studies in recent years, "We are not even close to a well-developed theory of policy implementation."[20]

Meanwhile, helpful hints, such as checklists intended to strengthen the hands of implementers, have been little more than laundry lists of variables that implementers would do well to "take into account." For instance, effective implementation demands that a program be "based on a sound theory relating to changes in target group behavior," and that "technical advice and assistance should be provided, and superiors should rely on positive and negative sanctions."[21] The development of practical insights into "what to do about" implementation-sensitive policy initiatives has thus remained arrested.

IMPROVING IMPLEMENTATION

Remedies for policies that suffer, or are at risk of suffering, from implementation sensitivity take three main forms.

Prevention

One school of thought contends that because ambitious public policies inevitably succumb to "goal displacement" in the course of implementation, public policymakers should curb their enthusiasm for complex interventions.[22] The recent intellectual foundations of this argument were laid by scholars who ventured into the field to study the progress of Great Society programs, only to discover that these innovations sometimes went sharply off course on the journey from Washington to their intended beneficiaries. Jeffrey Pressman and Aaron Wildavsky, for example, famously mapped the detours and distortions encountered in Oakland, California, by federal funds to train unemployed residents for good jobs as multiple uncoordinated agencies got into and disrupted the act.[23] Jerome Murphy found that federal grants under Title I of the Elementary and Secondary Education Act of 1965, intended to support innovations in schools with many disadvantaged students, were sometimes diverted instead to the general-purpose budgets of the recipient schools.[24] And in a book often taken to be an authoritative account of the community action programs of the federal Office of Economic Opportunity, Daniel Patrick Moynihan recounted how federal requirements for "maximum citizen participation" in the allocation of funds ended up subsidizing agitators determined to fight city hall, not work with it, thus fostering "maximum feasible misunderstanding."[25]

These authors did not leap from their findings to a general conclusion that federal interventions are doomed to fail, but their work (along with that of other researchers) was taken as supporting evidence by a growing cadre of neoconservatives who, in the journal *The Public Interest* and in other outlets, argued that an incompetent federal government was throwing money at problems it had no clue how to solve. In the 1970s this indictment became a prime element of conventional wisdom on the Right, in time reaching its apotheosis in Ronald Reagan's allegation that government is not the solution to the nation's problems but rather is itself the problem.

Although it is inarguable that nonexistent programs will escape problems of implementation, this anti-interventionist stance presents problems of its own. Inoculating programs against implementation sensitivity by not undertaking them in the first place is a defeatist approach that might be justified if the disquieting findings of the studies of implementation of certain Great Society programs proved they were indeed the misshapen progeny of Big Government that the neoconservative pundits took them to be, but there are other ways to interpret the record. For one thing, although case studies are indispensable for understanding how and why programs do or do not "work," single cases are less instructive for lesson drawing than are studies that compare programs across several sites, and these are rare in this literature. (Malcolm Goggin, in an important exception, examines the implementation of two child health care programs over time in several states.[26]) And some highly influential studies did not dig very deeply. Moynihan's evisceration of the community action program, for example, deployed a broad polemical brush that indelibly tarred the universe of those agencies in the eyes of pundits and (some) policy analysts.

Moreover, it may be that Great Society gambits in (for instance) workforce training and educational innovation for the disadvantaged veered off course not because they were doomed in advance by the herculean labors of implementation but because their federal designers failed to take their implementation sensitivity sufficiently into account on the front end. By the mid 1960s, when these programs were concocted, it was no mystery that workforce training involves federal, state, and local agencies; local training programs; and private firms in putative "networks" that are complex and difficult to steer and sustain. In 1955, for instance, Peter M. Blau, drawing on data collected in a state employment agency in 1948–1949, explored how the "different disturbances that constantly arose led to new patterns that superseded old ones" as "officials in active contact with clients redefined abstract procedures in terms of the exigencies of the situation and the dominant objectives of their task."[27]

Such "lessons," it seems, failed to register. Almost forty years after Blau's study (and two decades after the appearance of Pressman and Wildavsky's classic), Richard Nathan documented the difficulties of making federal "workfare" policies work and called both for greater attention to the challenges of implementation and for "dual implementation strategies" that honored the need "on the one hand for political leadership in the implementation process and on the other for taking managerial considerations into account in policy formulation."[28]

Nor was it then (or since) a revelation that local (and state) school systems tend to be introverted institutions, sometimes weakly if at all accountable to local government, that might be unable or unwilling to launch innovations that (more or less by definition) were unfamiliar and disruptive of standard practices. The overarching conclusion of empirical research on policy implementation, wrote Milbrey Wallin McLaughlin, is that "it is incredibly hard to make something happen, most especially across layers of government and institutions."[29]

Greater sensitivity by policymakers to the implications of implementation sensitivity might have increased the odds of success of programs that aimed at worthy goals but would likely not have seen the light of day without the authority, encouragement, and funding of the federal government. In seizing so avidly on implementation problems as a rationale for retrenchment of public sector programs, it would seem that neoconservatives were "mugged" (to recall Irving Kristol's famous bon mot) less by reality than by ideological animus.

Preemption

Unlike the Right, which sees fumbling implementation as evidence of overreaching by a witless government and therefore as an argument that the public sector should trim its sails, the Left inclines to view it as a source of disparities and inequities, the correction of which requires that government try harder to preempt the problems to which implementation sensitivity gives rise, both by clarifying the aims of policy and by expanding the scope of policies to close gaps that emerge as they are implemented.

Theodore Lowi, for example, argued that lack of specificity in policies opens the door to "interest group liberalism"—the parceling out of policy among special interests—which could be ameliorated if courts enforced "juridical democracy" by insisting that legislators make their handiwork more precise.[30] One much-noted caveat is that a degree of statutory ambiguity (perhaps most likely to arise on precisely the most politically sensitive issues) is often a price paid for successful coalition building, in which case the choice may lie between ambiguity and inaction.[31] A second rejoinder holds that too much legislative specificity can do more harm than good by tying too tightly the hands of those who seek to adjust policy to fit unforeseen and refractory situations in the course of implementation.[32]

Recently, prominent analysts of the politics of policy, including Theda Skocpol and Andrea Louise Campbell have emphasized the unfortunate and damaging consequences of "targeted" programs, the categories of which leave gaping holes into which needy citizens fall.[33] In a brilliant rendering of what implementation sensitivity meant in one salient case (that of her disabled sister-in-law, who sought to enroll in public programs in California after a car accident), Campbell recommends modeling social assistance programs on successful universalistic programs such as public education, Social Security, and (presumably) Medicare.[34]

Although universalistic programs are sometimes touted as an antidote to subnational variation and the vagaries of implementation, the term has no single meaning. "Universal" programs may encompass (1) the entire population as a right of citizenship (for instance, national health insurance in Western nations other than the United States); (2) everyone within a given category (for instance, demographic groups, as in Medicare and Social Security for the aged), or (3) all who meet specified criteria of need (such as Social Security Disability Insurance for the disabled). The one thing these three variants have in common is the absence of means testing (that is, eligibility based on income status).

The rub, of course, is that the United States has little taste for rights beyond the formal legal rights of individuals and resists entitlements to social protections, so the supply of universalistic programs is sharply constrained. Public education, sometimes invoked as a glowing exception, is in fact less exception than rule. Universal in principle—all American children have a right to free public education from kindergarten through grade 12—the right is in practice highly sensitive to how it is implemented. Because funds for public schools depend heavily on local property taxes, the quality of education that children receive is importantly—perhaps overwhelmingly—a function of the ability of their parents to afford housing in communities with good schools or to pay for private schooling.

Efforts such as the above-mentioned Title I of the Elementary and Secondary Education Assistance Act of 1965 and numerous variations in and around it, which aim to make educational opportunities more equitable, have generated some of the most revealing literature on implementation sensitivity and continue to do so. For instance, the No Child Left Behind program of 2001—"the single greatest expansion of the federal role in education since the original 1965 act"—left discretion to the states to adopt "arbitrary differences in statistical formulae" and to win waivers "idiosyncratic to each state," thus illustrating how "subtle differences in policy implementation may cause dramatic differences in measured outcomes" and generate "fifty ways to leave a child behind."[35]

Unlike societies that regard health care and various other benefits as rights the public sector should secure as a matter of equity, the United States envisions equity in *policy* as mostly a matter of fairness to hard-working taxpayers, whose earnings public programs can legitimately redistribute only to fellow citizens who supply rigorous proof that they are not only needy but also deserving. The determination to honor narrowly defined notions of the deserving poor (as "American as a sawed-off shotgun," to lift a phrase from Dorothy Parker) runs like an indelible national signature though Aid to Families with Dependent Children; Temporary Assistance to Needy Families, or TANF; the Supplemental Nutrition Assistance Program, or SNAP (i.e., food stamps); Medicaid; and other programs of social assistance—so much so that Jamila Michener calls it "the most critical component of American poverty governance."[36] In the nature of the case such programs leave substantial discretion to state and local implementers who interpret eligibility and other crucial questions in the mixed and murky context of federal laws and rules, regional and local norms, agency perspectives and practices, and the force and direction of whatever political winds happen to be blowing at the time.

In vast swaths of U.S. social policy, delegation by federal to state and local officials interposes between formulation and implementation[37]—a handoff in which federalism is seldom frictionless. A manager of the federal food stamp program explained to Michael Lewis that "where you live in this country makes a huge difference if you are poor," citing states with "sixty- or seventy-page documents people have to fill out to get benefits," and lamenting that "poor people are easy to wear down." Lewis cited arresting cases in point, for example: the Wyoming legislator who boasted of "how badly he had gummed up the state's nutrition programs" because "we pride ourselves on doing the minimum required by the federal government"; the Arizona congressman who proposed that the card used by those who get food stamp benefits "be made prison orange, conferring not just nutrition but shame"; and the effort by North Carolina to distribute federal disaster relief food benefit cards to residents of several severely flooded counties on Election Day in 2016, "to give poor people a choice between eating and voting."[38]

Another instructive case in point is MaxEnroll, a program by means of which the Robert Wood Johnson Foundation sought to boost enrollment in eight states of children who, under federal and state policies, were eligible for Medicaid and the Children's Health Insurance Program, or CHIP. The capacity of Medicaid and CHIP administrators within these states to augment enrollment depended on myriad "details"—for instance, the length and complexity of application forms; the availability of assistance for non–English speaking applicants; the diligence of outreach efforts, first to organizations (such as churches and community organizations) that might help identify eligibles and then to eligibles themselves; the efficiency and interoperability of Medicaid and other agencies' computer systems; and the timely handling of reenrollment before (and after) deadlines passed.

Political challenges compounded the bureaucratic. Efforts to persuade local social services officials that strict gatekeeping of enrollment should yield to a "culture of coverage" sometimes collapsed under the weight of media exposés of numbers—typically tiny—of Medicaid beneficiaries who did not meet criteria of eligibility for the full term of their enrollment, followed by indignant legislative investigations, executive embarrassment, and a renewed tightening of enrollment criteria. (This account of MaxEnroll comes from unpublished research by Richard

Nathan and this author.) The road to equity—foggy and potholed at best—may run not through quixotic efforts to create universalist programs but rather though fuller understanding of the workings of local cultures, organizational dynamics, and political championship.

Profit

Given the preponderance of economists among policy analysts, it is no surprise that "fixing the incentives"—manipulating the material interests of individuals, organizations, and, indeed, whole systems—is a popular prescription (and one seemingly more centrist than the two discussed above) for better aligning the aims of policies with their implementation. One highly touted variant is pay-for-performance, wherein policymakers identify desirable processes (screening for specified medical conditions, say) and give extra money to providers who carry them out. The approach has been applied mainly to individual providers, and its success is far from clear. One recent discussion finds pay-for-performance to be "highly contestable and contested" for several reasons: incentives of sufficient magnitude can be quite costly; the approach may inhibit "cooperative tendencies" among its objects; it is vulnerable to gaming; and, far from simple and self-executing, it works best when those participating in it have a hand in setting the rules of the game.[39]

When the challenge is to keep implementation on track throughout multiple levels of a complex formal organization, or a system of such organizations, as in the case of workforce policy and active living, the logic of pay-for-performance grows cloudier still. The much-heralded correct incentives, promulgated by policymakers from on high, often succumb to organizational refractions that render them less a reliable solution to implementation sensitivity than a part of the problem itself.

For example, officials in charge of training and referring unemployed people might be paid more to avoid cherry-picking easy cases and instead to persist with hard ones, but, as with any risk adjustment methodology, the hardness (hence, costliness) of a case is not easily disentangled from the efficiency of the management by which it is handled—a problem that in turn complicates the calibration of budgets in light of outcomes. In the case of active living, backing proposed changes in the built environment with bigger budgets—for more bike lanes, for instance—would surely boost incentives for those who are inclined to craft initiatives, but it is not clear how bigger budgets per se would solve the key implementation issues—getting agreement from communities, organizational staff, and political leaders about what constitutes well-designed lanes and where they should go. Safe Routes to School offers another cautionary tale. Many school districts were slow to tap available federal funds because parents feared "stranger danger"; teachers and principals were preoccupied with other matters (educating students, for instance); and political leaders (superintendents and school boards) at the top of school systems were reluctant to commit political capital to a seemingly extraneous endeavor.

Economic wisdom notwithstanding, incentives cannot make everything right.[40] Changing them may (or may not) be *necessary* to improve the implementation of programs, but doing so is seldom *sufficient* to keep implementation-sensitive policies on course as incentives filter through and get dislodged and disoriented by forces of culture, organization, and politics.[41]

A variant on the "let the money do the leading" theme contends that implementation will improve if public authorities, saddled with oppressive sources of inefficiency, such as cumbersome hiring rules (which are said to be inherent in their identity as *public* actors) contract out or otherwise rely for implementation on organizations in the private or nonprofit sectors. This approach, often dubbed New Public Management, may appeal if one concurs that specifically *public* features of implementation are the main reasons why the process falters. If, however, the sources of bumbling are less sectoral (public versus private) than organizational (such as the definition of role and mission[42]); the adequacy of problem solving tools; the influence of actors in the environment; and other variables common to and problematic for both sectors), it looks depressingly simplistic.[43] Moreover, public sector contracting-out does not solve—indeed, serves mainly to aggravate—the problem of organizational intelligence; that is, the contractors' capacity to know whether, in fact, the contracted entity is behaving as expected, given the difficulties of monitoring performance from a distance and the omnipresent temptation to game quantitative measures of outcome.

A third version of this strategy looks to competition—especially the push for profits or revenues by providers who compete within publicly set and managed rules of the game—to bring implementation into line with expectations. Health policy offers an illuminating case in point, namely, "managed" competition, which claims a unique ability to secure proper access to good quality health care at prices rising more slowly than those in less integrated systems.[44] Organizations that are instructed and "incentivized" to compete may, however, decline to do so or may temporize. After the British NHS introduced new competitive pressures in 1991, for example, the presumed competitors sometimes opted to "sacrifice organizational interests in favour of a perceived greater good . . . Managers within Primary Care Trusts . . . whilst clear about the need to balance their books, were not inclined to do this at the expense of other local organizations [and] were reluctant to exercise autonomy because they did not want to 'destabilize' other local organizations."[45]

Moreover, even when the managed competitors do more or less as they are told, the record to date, which rests mainly on studies of implementation in the Netherlands and Switzerland, suggests that there is no easy exit from the complications that bedevil active living and other implementation-sensitive programs.[46] The beneficial effects of managed competitive incentives depend importantly on culture (do subscribers in the competing insurance plans trust them to select high-quality rather than merely lower-cost providers?), organizational dynamics (do insurers seek to circumvent government rules that preclude cherry-picking of less risky subscribers?), and political leadership (are policymakers willing to release competing insurers from a host of regulatory shackles? Can they effectively monitor what transpires in these competitive markets and intervene to fix problems?).[47] Similar issues arise for a range of policy arenas, including education, in which market panaceas include school choice, charters, and vouchers.[48]

IMPLEMENTATION REVISITED

The analytical and prescriptive frustrations that afflict the study and improvement of implementation are, at base, products of the ways in which the inquiry has been framed—namely, as an empirical test of a normative (functional) model. As noted above, democratic theory postulates a sequence of decision-making steps over time: elected officials enact laws that set goals; the laws are entrusted to bureaucratic specialists who refine them into regulations; the fleshed-out policy initiative goes to federal, state, and local actors whose job is to implement it in ways that conform to the designated ends and means that have been crafted by their superiors at higher levels of policymaking. The policy process is thus a sequence of hand-offs ending in the hands of "implementers."

It is of course difficult to demarcate precisely the point at which policy shades into implementation, but that there *is* some such point is integral to the functionalist model and to the "lasting functionality" that the dichotomy between policy/politics and administration/implementation—the latter viewed as "a presupposed residual in goal achievement"[49]—enjoys in democratic theory.[50] Implementation that is "incongruent"—inconsistent or at variance with the goals of policy—invites reproach and a merry chase to uncover both the sources of failure and the most promising repairs.[51] Alas, the chase mainly discloses that sources are many and uncertain and that correctives are banal and unreliable.

According to Peter L. Hupe, the variables that potentially explain implementation are "almost endless."[52] One hardy perennial is of course "goal clarity" and its obverse, "goal ambiguity," but goals emerging from political negotiations may be ambiguous in different ways, leaving them "susceptible to multiple interpretations."[53] The contexts in which implementation takes place also differ "endlessly."[54] "Unforeseen contingencies" enter the picture, as do "specific characteristics of the players and the institutional settings in which they operate."[55] "Professional behavior and the use of common sense" loom large among these characteristics.[56] Hierarchies and layers of governance within and across institutions cannot be discounted, nor can "governance skills" such as "managerial competence and other forms of craftsmanship."[57]

This far-flung intellectual terrain turns policy into "an interwoven somewhat" and leaves unclear when and whether one is researching rule setting (formulation) or rule application (implementation).[58] Answers can only be, as Hupe puts it, "a matter of empirical observation on the basis of operationalised theoretical concepts."[59] In plainer terms, implementation is an exercise in improvisation concocted in the course of interpretation of the exigencies and opportunities posed in and by shifting and contingent contexts. Or as Angela Browne and Aaron Wildavsky wrote in a similar vein in 1984, implementation "is no longer solely about getting what you once wanted, but, instead, it is about what you since learned to prefer until, of course, you change your mind again."[60] Implementation, it seems, is in essence a sense-making activity that by its very nature defies capture in general and theoretical cages.

In short, implementation studies need to be liberated from a malingering functional model and viewed through refreshed empirical lenses. The first step is to bracket (put temporarily out of consideration) hoary democratic axioms, the Wilsonian dichotomy, Weberian images of bureaucratic autonomy, and the voluminous learned disquisitions on the pros and cons of the functional model and to adopt instead what might be called a phenomenological approach. Instead of canvassing the spatial-temporal requisites of democratic fidelity, this orientation gazes simply and directly at the contours and contents of implementation projects—the "micromechanics of power"[61]—in specific local sites. Doing so suggests that active living policy is not some neatly defined species but rather bits and pieces of initiatives and interventions in which policy and implementation mingle—and sometimes merge. Four patterns ensue.

Explicit Policy, Active Implementation. First, in all five cities studied here, one does find active living promoted by policy enactments (laws, rules) and initiatives (grants from foundations) proffered from above to implementers below. Examples include CDC awards, provisions in sections of federal transportation laws, and the ALbD program. Such policies can also be generated by cities themselves—for instance, Complete Streets ordinances. By no means self-enforcing, the fate of such initiatives is heavily implementation-sensitive—that is, dependent (as the previous chapters have indicated) on contingent forces of local particularism, intersectoral pluralism, and political championship.

Explicit Policy, Inactive Implementation. Explicit policies are not always activated, a source of frustration surrounding programs that sit fruitlessly "on the books" for want of effective implementation. In most of the years covered by this study, for example, the federal government offered money to applicants for Safe Routes to School projects—money that was in some cases not pursued (or not much deployed) because parents, teachers, superintendents, and school boards could not agree on what, if anything, they wanted to do with it. Another case in point is funding for a gym in a lower-income neighborhood in Louisville, which was left in limbo because the multiple voices within the community could not decide how to design the facility.

Implicit Policy, Active Implementation. Each site lies under the authority of laws, regulations, codes, and programs that are not overtly aimed at promoting active living but can be enlisted for that purpose. In Louisville, for instance, a local activist diligently combed zoning codes and ordinances in search of provisions that could be interpreted as permitting the construction of a sidewalk on a busy road. Cycling advocates may seek to define road repaving projects as an occasion for the addition of painted lines that designate bike lanes. Arguably, in such cases implementation is the prime mover that transforms public policies into projects that are both innovative and consistent with "intent." From another perspective, implementation here merges with policy by making manifest potential uses that lie latent within the provisions in question.

Implicit Policies, Inactive Implementation. This cell may be the most heavily populated of the four, but no accurate census is possible because its contents by nature fall within the misty domain of what might have been. For instance, plans for altering traffic patterns in downtown Wilkes-Barre would seem to offer an opening for bike lanes, but trial balloons popped, and the documentation and advocacy required to reinflate them lay beyond the organizational and financial resources of the advocates.

In sum, implementation (and its sensitivity) may be viewed as patterns in which policies, both explicit and implicit, constitute enabling contexts that in turn may be interpreted as entrepreneurial openings by innovators who set about improvising new projects and programs.

IMPLEMENTATION AS SENSE-MAKING

However ardent an analyst's yearning to be constructive, the workings of implementation-sensitive programs cannot be systematically improved because in the nature of the case such programs defy systematization. These programs are not only profoundly contingent on the cognitive orientations and social networks of implementers but also shaped by accidents, happenstance, luck—and by all the indeterminacies that surface.[62] Contingency triggers endless frustration among policy analysts because it defies the scientific pretensions of generalization, replicability, predictability, modeling, and the rigorous apportioning of causal influence among variables. As explained in the previous chapter, one cannot generalize about how "it" works because "it" works differently across both space and time. The best one can do analytically, therefore, is to rely on "sensitizing concepts," "metis," "sense-making," and "thick description"—forms of practical knowledge, the utility of which in concrete cases (not to mention as a foundation for scientific aspirations) is as contingent as implementation itself.[63]

The active living cases offered in this study suggest that making sense of implementation entails three basic tasks for practitioners and analysts alike.

Contextualization

Because "structural changes of this kind affect and are shaped by both the physical (built) and social environment, and are therefore inherently contextual and complex," sense-making starts with a close and careful descriptive account of the setting and situation in which implementation is to proceed.[64] This exercise depends (in Clifford Geertz's terms) on "thick" description, up-close and personal insights, and "local knowledge" that yields a serviceable grasp of "the way things work around here."[65] As a species of policy learning, contextualization is a matter of learning *that*, which is to say, of developing a deep-reaching (although necessarily incomplete) factual grasp of the "character" of the site(s) in question. (In an excellent exploration of why "policies that do not address . . . organizational, professional and social contexts are unlikely to achieve successful implementation," Susan Watt and colleagues examine the contrasting outcomes of two programs in Ontario, one that allowed new mothers a hospital stay of up to sixty hours and another that offered follow-up home visits to mothers and infants by public health officials.[66])

"Knowledgeable individuals exercise judgment within their domain of action because (and to the extent that) they have successfully completed a period of socialization (sociocultural, professional, organizational, and usually a combination of all these) that has enabled them to appreciate and take account of subtle aspects of context when making distinctions."[67] In the cases examined here, the exploration of context includes digging into history (previous attempts to promote active living), cultural orientations toward the built environment (in city hall and in neighborhoods), organizational perspectives (which agencies, and which officials within them, have responsibility for which facets of the built environment), and political power (the present and potential ability and will to move agenda items into action) of mayors, council representatives, county officials, governors, state legislators, and federal officials.

This counsel may seem elementary—indeed, gratuitous—but it departs sharply from strategies often adopted both by funders in government and in foundations who seek to gauge the capacity of applicant communities to put policies successfully into practice and by locals who seek expertise to help them correct the wayward course of implementation. A common approach to contextualization assembles lists of objective indicators—demographic shifts into and within a region or community, number and condition of sidewalks and bike lanes, and surveys of who exercises how many times per week, for example—and mistakenly assumes both that these data speak for themselves and that they are not only necessary but largely sufficient to guide the work of program managers. Contextual analysis seeks "to understand the rules of the game well enough to know the standard moves and have a repertoire of effective countermoves" but the usual data, alas, often fail to illuminate the cultural, organizational, and political forces that lie at the heart of implementation.[68]

Sponsors and funders then may proceed to confirm that a little knowledge is a dangerous thing by supplementing statistical profiles of promising places with site visits in which teams of advisors whose members typically have but a glancing acquaintance with the sites in question endure one to two day-long dog-and-pony shows in which leading lights in local power structures expound on their visionary plans and attest to their stellar capacities for implementation. Money in hand, the local leaders return to their packed to-do lists and delegate the project to middle-level officials, leaving funders, who have shifted attention to the creation of new programs and the awarding of new grants, to wonder a year or so later why progress in these previously funded projects has stalled. Then they compound the problem by dispatching to the sites consultants, self-proclaimed connoisseurs of technical assistance, who arrive with prefabricated stratagems that they supplement (or jettison) on the spot with "insights" garnered from the local actors they are handsomely paid to assist.[69]

Interpretation

The second step in sense-making for implementation is an explanatory exercise that scrutinizes "context clues" for the practical meaning behind or within the situational knowledge acquired by means of thick description, thus moving beyond learning *that* to learning *why*.[70] The aim, harking back to the Chicago school of sociology and W. Thomas's "elementary program," is "to resort to the 'definition of the situation,' that is, to understand the perspective in which actors see the situation in which they find themselves."[71]

Why did prior attempts to implement proposals, projects, and programs (or components of them) within a jurisdiction advance or stagnate, and what do plausible explanations imply for current and future plans? In the case of active living, one seeks to explain why, for instance, one neighborhood or another was accepting of or resistant to the pedestrian- and cyclist-friendly changes in the built environment proposed in the past; why attempts to sell active living to leaders in the Department of Health or bike lanes to highway engineers in the Departments of Transportation or Public Works turned out as they did; and why mayors or members of the city council ignored, demurred, gave lip service to, or embraced such initiatives. By carefully "reading" factual information with an eye to discerning parallels between past and present (always an imprecise exercise at best), interpreters try to flag detours and entry points in hopes that such explanatory exegeses enable implementers to ask reasonable questions and to contemplate critically the answers put before them by stakeholders and experts alike. Moreover, because politics is "driven by how people interpret information [and] much political activity is an effort to control interpretations," the analytics of interpretation demand not only an explanatory account of the processes and outcomes of policy but also of the political deployment of interpretations, and of their practical effects, in the policy process.[72]

Improvisation

Contextualization and interpretation—discovering that and exploring why policies have evolved as they did in the course of implementation in one or more venues—might seem to be not only a necessary preliminary to but also perhaps a sufficient foundation for predicting what approaches will work and for guiding programs effectively into place. This optimism is misguided, however, because the *how to do it* stage of policy learning does not directly follow from the two prior stages and, indeed, is not a teachable or learnable skill (in the conventional sense) at all. Because this "stage" is less a matter of learning or puzzle solving than it is of "groping"—searching for more or less secure footing on more or less spongy terrain—the best one can achieve is a "perfectly theorized manner of having no theory."[73]

Managing implementation-sensitive programs demands improvisation—adaptive, make-it-up-as-one-goes-along best guesswork, illuminated by the fruits of contextualization and interpretation, to be sure, but never proved or validated by the rigorous testing of hypotheses. Improvisation is how implementers cope with uncertainties and contingencies, and, by definition, these unsettling conditions abound in implementation-sensitive programs.

Improvisation, however, is not some managerial equivalent of free verse. Its essence is discretion within rules, freedom within structure—a product of interpretive deliberations on the meaning(s) of multiple contexts in an environment oftentimes marked by "complexity, indeterminacy, and the necessity to act on the situation at hand."[74] These deliberations, as Hendrik Wagenaar explains in a valuable study of a Dutch immigration official in a Department of Implementation Policy, combine "hidden, taken-for-granted routines; the almost unthinking action, tacit knowledge, fleeting interactions, practical judgments, self-evident understandings and background knowledge, shared meanings, and personal feelings that constitute the core of administrative work."[75]

Because no set of cultural understandings supplies, as Howard Becker put it, "a perfectly applicable solution to any problem people have to solve in the course of their day," they therefore must "remake those solutions, adapt their understandings to the new situation in light of what is different about it."[76] These remakings and adaptings seek a "joint construction . . . of shared meanings" in which the participants "feel their way together, like [jazz] musicians, looking for the joint actions that will produce a result more or less acceptable to all."[77] Becker illustrates the interplay between convention and contingency with a "complication," the "action logic" of which applies as well to public agencies (say to Departments of Health and Transportation) as to individual agents—namely, a divorced woman with two young children and a lover. "In this case, the couple's freedom of action is limited, and there is no cultural model that suggests what they should do to resolve the difficulties. . . . The models that serve to form couples and those that serve to raise children suggest incompatible solutions, and the partners are forced to impose something. They have to improvise."[78]

So, too, must would-be change agents seeking to implement innovations. Matthew Stewart's musing on successful strategizing in the corporate world—that such strategies "emerge through action, [through a] play-it-by-ear kind of strategy-making . . . [that] stems from a healthy recognition that the world is generally too complex for our simple plans"—extends beyond the private sector.[79] For example, proposing that federal programs can be viewed as, in essence, a lawn party—"the occasion for a gathering, [at which] the guests do pretty much as they please"—Eleanor Farrar and colleagues found in the Experience-Based Career Education program a pattern of evolution from "particular configuration[s] of interpretations" underpinning local improvisations that varied with the nature of the particular problems school systems chose to address.[80] In similar vein, James Spillane and colleagues contend that in the education arena, "understandings and beliefs about instruction, subject matter, and the like are worked out in the context of instructional practice," wherein implementing agents must use "their prior knowledge and experience to notice, make sense of, interpret, and react to incoming stimuli—all the while actively constructing meaning from their interactions with the environment of which policy is part."[81]

In the active living cases examined in this study, one continually discovers crucial "you can't plan that" contributions to the sites' progress. Louisville benefited from the serendipitous coincidence of two cycling enthusiasts in the mayor's chair over many years plus the presence in the Department of Public Works of a dedicated cyclist with a master's in public health and impressive political skills. Like Wilkes-Barre, many Rust Belt cities with modest populations are financially beleaguered and physically battered. That city, however, combined a long-running association of downtown businesses that enjoyed astute and indefatigable leadership and a mayor who concurred that a walkable center city—one in which residents, shoppers, workers, and tourists wanted to stroll around and felt safe when doing so—was essential to the city's financial and cultural rehabilitation. In Albuquerque plans for Complete Streets and other enhancements for active living advanced under the patient prodding of a dedicated city council member and through networks of personal connections that enlisted advocates and ALbD grantees, sympathetic aides to elected officials, and young planners at MRCOG, each and all determined to leverage their individual and collective connections with political leaders at the city and county levels. In Sacramento an avid cyclist who was determined to build support for Safe Routes to School within the school system also happened to be a first-rate navigator of the challenging politics required to do so. And New York's sizable portfolio of active living efforts reflected the twelve-year presence in city hall of a mayor who made gains in public health his top priority instead of, for instance, reducing crime, the main preoccupation of his predecessor.

Improvisation is an interpretive exploration in discerning how enabling contexts may give rise to entrepreneurial openings for policy departures. Successful improvisation depends on both luck and judgment, the latter being what A. J. Liebling called "the sum of experience and flair."[82] The crucial elements are not skills that can be learned or taught. The quest for generalizations and predictive models, derived from replicable statistical tests that apportion causal weight among quantifiable variables and transcend the idiosyncrasies and contingencies of individual cases creates expectations that cannot be fulfilled, no matter how ingenious the methodological tortures inflicted on the body politic. The best policy can do (and the most that sound policy analysis attempts) is to furnish aspiring implementers with rich food for thought, bits and pieces of illuminating information, and interpretations of the raw materials of sociopolitical life that may—or may not—make sense of and for the improvisational policymaking that implementation demands and rewards.

NOTES

INTRODUCTION

Epigraph: Charles Sanders Peirce, *The Collected Works of Charles Sanders Pierce*, ed. Charles Hartshorne and Paul Weiss, vol. 8, *Reviews, Correspondence, and Bibliography*, sec. 12 (Cambridge, Mass.: Harvard University Press, 1958).

1. M. Katherine Kraft and Lawrence D. Brown, "Active Living by Design as a Political Project: Challenges at Three Levels," *American Journal of Preventive Medicine* 37, no. 6 (2009): 453–54.
2. Richard Neustadt, *Presidential Power and the Modern Presidents: The Politics of Leadership* (New York: Wiley, 1960), 9.
3. Daryl Siedentop, *Introduction to Physical Education, Fitness, and Sports*, 7th ed. (New York: McGraw-Hill Higher Education, 2009), 24–25.
4. Helena Rosenblatt, *The Lost History of Liberalism: From Ancient Rome to the Twenty-First Century* (Princeton, N.J.: Princeton University Press, 2018), 220.
5. William H. McNeill, *Keeping Together in Time: Dance and Drill in Human History* (Cambridge, Mass.: Harvard University Press, 1995), 140.
6. Siedentop, *Introduction to Physical Education*, 46.
7. Heather Maxey, Sandra Bishop-Josef, and Ben Goodman, *Unhealthy and Unprepared: National Security Depends on Promoting Healthy Life-Styles from an Early Age* (Washington, D.C.: Council for a Strong America, October 2018).

8. Johan Huizinga, *Homo Ludens: A Study of the Play Element in Culture* (Boston: Beacon, 1950), 3, 197.
9. McNeill, *Keeping Together*, 7, 42, 138, 150.
10. Marcel Proust, *A la recherche du temps perdu: Le temps retrouve* (Paris: Gallimard, 1989), 5.
11. Joseph Rykwert, *The Seduction of Place: The History and Future of the City* (New York: Vintage, 2002), 175.
12. Jack W. Berryman, "Motion and Rest: Galen on Exercise and Health," *Lancet* 380, no. 9838 (2012): 210–11.
13. John MacDonald, Charles Branas, and Robert Stokes, *Charging Places: The Science and Art of New Urban Planning* (Princeton, N.J.: Princeton University Press, 2019), 128.
14. Sander Gilman, *Franz Kafka: The Jewish Patient* (New York: Routledge, 1995), 106.
15. Peter Gay, *Weimar Culture: The Outsider as Insider* (New York: Norton, 2001), 77; and Benjamin Carter Hett, *The Death of Democracy: Hitler's Rise to Power and the Downfall of the Weimar Republic* (New York: St. Martin's Griffin, 2018), 99.
16. John O'Hara, *Appointment in Samara* (New York: Vintage, 1934), 204.
17. Centers for Disease Control and Prevention and the American College of Sports Medicine, "A Recommendation," *Journal of the American Medical Association* 273, (1995): 402–7.
18. United States Department of Health and Human Services, *Physical Activity and Health: A Report of the U.S. Surgeon General: Executive Summary* (Washington, D.C.: U.S. Department of Health and Human Services, 1996).
19. United States Department of Health and Human Services, *2008 Physical Activity Guidelines for Americans: Be Active, Healthy, and Happy!* (Washington, D.C.: U.S. Department of Health and Human Services, 2008).
20. Harold W. Kohl III and Heather Cook, eds. *Educating the Student Body: Taking Physical Activity and Physical Education to School* (Washington, D.C.: National Academies Press, October 30, 2013).
21. For sophisticated critiques of the anti-obesity crusade, see Paul Campos, *The Obesity Myth: Why America's Obsession with Weight Is Dangerous to Your Health* (New York: Gotham Books, 2004); Glenn Gaesser,

Big Fat Lies: The Truth About Your Weight and Your Health (Carlsbad, Calif.: Gurze, 2002); and Eric J. Oliver, *Fat Politics: The Real Story Behind America's Obesity Epidemic* (New York: Oxford University Press, 2006). See also the *New York Times* article by Andrew Pollack, which reports that the decision of the American Medical Association to designate obesity as a disease overrode the recommendation against doing so of its own Council on Science and Public Health. Andrew Pollack, "AMA Recognizes Obesity as a Disease," *New York Times*, June 19, 2013, N1.

22. Oliver, *Fat Politics*, 145.
23. Deborah Stone, *Policy Paradox: The Art of Political Decision Making*, 2nd ed. (New York: Norton, 1997), 1.
24. An excellent and wide-ranging source for evidence and its practical implications is the Active Living Research website, https://activelivingresearch.org, which was launched by the Robert Wood Johnson Foundation as part of its portfolio of active living initiatives. See also Bradley J. Cardinal and Kris Day, "Special Issue: Active Living Research," *American Journal of Health Promotion* 21 (2007): Supplement.
25. U.S. Department of Health and Human Services, *Physical Activity and Health*, 10.
26. National Center for Health Statistics, *Health, United States, 2012: With Special Feature on Emergency Care* (Hyattsville, MD: National Center for Health Statistics, 2013), tables 60 and 67.
27. David R. Bassett Jr., Holly R. Wyatt, Helen Thompson, John C. Peters, and James O. Hill, "Pedometer-Measured Physical Activity and Health Behaviors in United States Adults," *Medicine and Science in Sports and Exercise* 42, no. 19 (2010): 1819–25.
28. Kenneth Burke, *On Symbols and Society*, ed. Joseph Gusfield (Chicago: University of Chicago Press, 1989), 242.
29. Robert Wood Johnson Foundation, *Active Living by Design: An RWJF National Program* (Princeton, N.J.: RWJF, last updated February 4, 2013), 2.
30. RWJF, *Active Living*, 2.
31. For case studies in point, see the special issue of the *Journal of Health Politics, Policy and Law*, "Active Living, the Built Environment, and the Policy Agenda," vol. 33 (2008).

32. Adam Oliver and Lawrence D. Brown, "A Consideration of User Financial Incentives to Address Health Inequalities," *Journal of Health Politics, Policy and Law* 37, no. 2 (2012): 201–26.
33. Robert S. Lynd and Helen Merrell Lynd, quoted in Tom Lewis, *Divided Highways: Building the Interstate Highways, Transforming American Life* (New York: Penguin, 1997), 34; and Sam Bass Warner and Andrew Whittemore, *American Urban Form: A Representative History* (Cambridge, Mass.: MIT Press, 2012), 74–77.
34. Lewis, *Divided Highways*, 293, 284.
35. Tom Vanderbilt, *Traffic: Why We Drive the Way We Do* (New York: Vintage, 2009), 16.
36. Vanderbilt, *Traffic*, 16.
37. Sherry Everett Jones and Sarah Sliwa, "School Factors Associated with the Percentage of Students Who Walk or Bike to School, School Health Policies and Practices Study 2014," *Preventing Chronic Disease* 13 (2016): 1.
38. Herbert J. Gans, *The Urban Villagers: Group and Class in the Life of Italian-Americans* (New York: Free Press, 1962); Alan Enrenhalt, *The Lost City: The Forgotten Virtues of Community in America* (New York: Basic Books, 1995); and Jane Jacobs, *The Death and Life of Great American Cities* (New York: Random House, 1961).
39. Fran Tonkus, *Space, the City and Social Theory* (Walden, Mass.: Polity, 2005), 19.
40. Alan Altshuler and David Luberoff, *Mega-Projects: The Changing Politics of Urban Public Investment* (Washington, D.C.: Brookings Institution, 2003).
41. Lewis, *Divided Highways*, 184.
42. Lewis, *Divided Highways*, 118, 214, 241.
43. Eric Klinenberg, *Going Solo: The Extraordinary Rise and Surprising Appeal of Living Alone* (London: Penguin, 2012).
44. Ted Mann, "New PAC to Back Pedestrians, Bikes," *Wall Street Journal*, April 25, 2013, A17.
45. Ralph Buehler and John Pucher, "Urban Transport: Promoting Sustainability in Germany," in *Lessons from Europe? What Americans Can Learn from European Public Policies*, ed. R. Daniel Keleman (Los Angeles: Sage, 2014).

46. Buehler and Pucher, "Urban Transport," 140–43.
47. Buehler and Pucher, "Urban Transport," 144.
48. Buehler and Pucher, "Urban Transport," 146.
49. Buehler and Pucher, "Urban Transport," 145.
50. Altshuler and Luberoff, *Mega-Projects*, 1, 3.
51. Rykwert, *The Seduction of Place*, 245, 158.
52. Rykwert, *The Seduction of Place*, 5. For a fine account of the themes and phases of downtown development in the second half of the 20th century, see Carl Abbott, "Five Downtown Strategies: Policy Discourses and Downtown Planning Since 1945," in *Urban Public Policy: Historical Modes and Methods*, ed. Martin V. Melosi, 5–27 (University Park: Pennsylvania State Press, 1993).
53. Clifford Geertz, *Life Among the Anthros and Other Essays*, ed. Fred Inglis (Princeton, N.J.: Princeton University Press, 2010), 252.
54. Clifford Geertz, *The Interpretation of Cultures* (New York: Basic Books, 1973), 313.

1. WILKES-BARRE

1. Thomas M. Leighton, "I Believe" (speech, Wilkes-Barre, Pennsylvania, June 9, 2005).
2. Handout given to the author in the course of an interview: Diamond City Partnership of Wilkes-Barre, Pennsylvania, "Downtown Wilkes-Barre Profile," Pub. 1/15, January 2015.
3. Michele G. Schasberger, Carol S. Hussa, Michael F. Polgar, Julie A. McMonagle, Sharon J. Burke, and Andrew J. Gegaris Jr., "Promoting and Developing a Trail Network Across Suburban, Rural, and Urban Communities," *American Journal of Preventive Medicine* 37, no. 6 (December 2009): 336–44.
4. A review of fitness and related activities may be found in Joint Urban Studies Center, *Healthy Communities: Luzerne County Assessment* (Wilkes-Barre, Penn.: Joint Urban Studies Center, May 25, 2007).
5. Michele G. Schasberger, Jessica Raczkowki, Lawrence Newman, and Michael F. Polgar, "Using a Bicycle-Pedestrian Count to Assess Active Living in Downtown Wilkes-Barre," *American Journal of Preventive Medicine* 43 (2012): 399–402.

6. Schasberger et al. (2012), S400.
7. Joe Dolinsky, "Public Square Business Owners Look for Foot Traffic," *Wilkes-Barre Times Leader*, June 14, 2015, 2A.
8. See also Jennifer Learn-Andes, "Work Begins on River Street Re-Design," *Wilkes-Barre Times Leader*, June 16, 2015, 1A, 4A.
9. Edward Lewis, "Gun Violence Shoots Up 21 Percent in Wilkes-Barre," *Wilkes-Barre Times Leader*, July 11, 2013.
10. Handout given to the author in the course of an interview: Diamond City Partnership of Wilkes-Barre, Pennsylvania, "Eight Key Takeaways from DCP's 2014 Downtown Perception and Use Survey" (2014).
11. James Q. Wilson, *Political Organizations* (New York: Basic Books, 1973).
12. For a succinct summary of the "trail planning and construction process in Wyoming Valley," see Schasberger et al., "Promoting and Developing a Trail Network," S342, table 3.
13. "Casey and Kanjorski Announce $950,000 in Federal Funding for Renewal of Coal Street Park in Wilkes-Barre," May 18, 2009, Senator Bob Casey website, casey.senate.gov/news/releases/casey-and-kanjorski-announce-950-000-in-federal-funding-for-renewal-of-coal-street-park-in-wilkes-barre.

2. LOUISVILLE

1. For an overview of the ALbD project, see Nina L. Walfoort, Jennifer J. Clark, Marigny J. Bostock, and Kathleen O'Neil, "Active Louisville: Incorporating Active Living Principles into Planning and Design," *American Journal of Preventive Medicine* 37 (2009): S368–76.
2. Louisville-Jefferson County Metro Government, Ordinance No. 128, Series 2019: An Ordinance Amending Ordinance 15, Series 2008, The Complete Streets Policy and Adopting the Policy as an Amendment to Plan 2040, The Comprehensive Plan (Case No. 19AREA-PLAN0001), https://louisvilleky.gov/document/completestreetsordinance1282019pdf.
3. Kentucky Youth Advocates, *Kentucky Complete Streets Toolkit for Advocates*, prepared by Andrea Plummer (2011), p. 2. http://kyyouth.org/wp-content/uploads/2013/09/KY-Complete-Streets-Toolkit.pdf.

4. Barbara McCann, "Complete the Streets for Smart Growth," *On Common Ground*, June 1, 2007, https://icma.org/documents/complete-streets-smart-growth, p. 27.
5. Walfoort et al., "Active Louisville," S371.
6. See the National Association of City Transportation Officials website, https://nacto.org/.
7. Louisville Metro Department of Economic Growth and Innovation, *Vision Louisville Phase 2 Report* (Louisville, Ky., December 2013), p. 110, https://louisvilleky.gov/document/vision-phase-2-reportpdf.
8. Branden Klayko, "Does Move Louisville Have Legs? A Critical Analysis of Louisville's 20-Year Transportation Plan," *Broken Sidewalk*, April 15, 2016.
9. Philip M. Bailey, "Fischer Unveils $1.4B Transportation Plan," *Louisville Courier-Journal*, April 14, 2016, https://www.courier-journal.com/story/news/politics/metro-government/2016/04/04/plan-citys-mobility-future-must-car-lite/82998494.
10. See the Bicycling for Louisville website, www.bicyclingforlouisville.org.
11. See the League of American Bicyclists website, www.bikeleague.org.
12. Ian Dille, "The 50 Best Bike Cities of 2016," *Bicycling*, September 19, 2016, www.bicycling.com/culture/news/the-50-best-bike-cities-of-2016.

3. ALBUQUERQUE

1. City of Albuquerque, "Prescription Trails," December 16, 2009, http://www.cabq.gov/parks/prescription-trails.
2. Joanne McEntire, quoted in a four-page document held by the author: Robert Wood Johnson Foundation, *Active Living by Design Community Partnership Profile: Albuquerque Alliance for Active Living*, n.d., p. 1.
3. Albuquerque Prescription Trails, "Walking Guide 2012," www.prescriptiontrailsnm.org.
4. For an overview, see Niraj Chokshi, "Population Growth in New Mexico Is Approaching Zero," *Washington Post*, January 17, 2014.
5. Complete Streets Ordinance, F/S O-14-27, City of Albuquerque Twenty-First Council, 2015, https://www.cabq.gov/council/documents

/councilor-district-2-documents/CompleteStreetsLegislationPacket .pdf.

6. City of Albuquerque, "Integrated Development Ordinance" (2017, as amended), http://www.cabq.gov/planning/codes-policies-regulations /integrated-development-ordinance.
7. See, for example, a nine-page document held by the author, Mid-Region Metropolitan Planning Organization (MRCOG), "Future 2040: Metropolitan Transportation Plan: Executive Summary," executive summary, April 2015.
8. James Brasuell, "Coming This Spring: Bikeshare Pilot Program in Albuquerque," *Planetizen*, February 28, 2015, https://www.planetizen .com/node/74567.

4. SACRAMENTO

1. United States Census Bureau, "QuickFacts Sacramento City, California," accessed December 28, 2020, https://www.census.gov/quickfacts /sacramentocitycalifornia. For other useful data see Joseph Parilla, Sifan Liu, and Marek Gootman, "Charting a Course to the Sacramento Region's Future Economic Prosperity," Brookings Institution, April 30, 2018. https://www.brookings.edu/wp-content/uploads/2018 /04/sacramentoregioneconomicprosperity_fullreport.pdf.
2. David Vogel, *California Greenin': How the Golden State Became an Environmental Leader* (Princeton, N.J.: Princeton University Press, 2018), 157–58.
3. State of California Department of Justice, California Environmental Quality Act (CEQA), accessed December 28, 2020, https://oag.ca.gov /environment/ceqa.
4. See Miriam Pawel, "What Makes California Politics So Special?" *New York Times*, August 19, 2018, 4.
5. Vogel, *California Greenin,'* 181–88.
6. U.S. Department of Transportation, "VMT Per Capita," accessed December 28, 2020, https://www.transportation.gov/mission/health /vmt-capita.
7. AB-2434 Strategic Growth Council: Health in All Policies Program, California Legislature 2017–2018 Regular Session (February 14, 2018),

https://leginfo.legislature.ca.gov/faces/billTextClient.xhtml?bill_id=201720180AB2434.

8. Safe Routes Partnership, accessed December 28, 2020, https://www.saferoutespartnership.org.
9. Jeffrey Pressman and Aaron Wildavsky, *Implementation: How Great Expectations in Washington Are Dashed in Oakland* (Berkeley: University of California Press, 1973).
10. Jeff Morales, "California Blueprint for Bicycling and Walking: A Report to the Legislature," Report to the Legislature, Prepared Pursuant to the Supplemental Report of the 2001 Budget Act, May 2002, https://dot.ca.gov/-/media/dot-media/programs/safety-programs/documents/ped-bike/f0018148-california-blueprint-bicycling-walking-report-a11y.pdf.
11. Sacramento Area Council of Governments, "Sacramento Region Blueprint," accessed December 28, 2020, https://www.sacog.org/sacramento-region-blueprint.
12. Mike McKeever, "Sacramento Region Develops New Paradigm for Transportation Planning," Prepared for Presentation to U.S. Senate Committee on Environment and Public Works, Field Briefing, California State Capitol, September 8, 2008, p. 7.
13. Sacramento Area Council of Governments, *2016 Metropolitan Transportation Plan/Sustainable Communities Strategy*, February 18, 2016, www.sacog.org/2016-mtpscs.
14. Sacramento County Board of Supervisors, *Sacramento County General Plan*, November 9, 2011, https://planning.saccounty.net/PlansandProjectsIn-Progress/Pages/GeneralPlan.aspx.
15. Alicia Brown, Project Coordinator, Walks Sacramento, Letter to Garrett Norman, Assistant Planner, City of Sacramento Community Development Department (December 8, 2017), https://www.walksacramento.org/wp-content/uploads/2018/01/Spanos-Natomas-Crossing-WALKSac-20171208-Letter.pdf.
16. Center for Active Design, "Active Design Guidelines," 2010, https://centerforactivedesign.org/guidelines.
17. Center for Active Design, "Sacramento Codifies Healthy Design at All Scales," accessed November 2, 2015, https://centerforactivedesign.org/sacramentoactivedesignpolicy.

18. City of West Sacramento, "Crime Prevention Through Environmental Design," accessed December 29, 2020, https://www.cityofwestsacramento.org/government/departments/police/crime-prevention-outreach/cpted.
19. Edie E. Zusman, Sara Jensen Carr, Judy Robinson, Olivia Kasirye, Bonnie Zell, William Jahmal Miller, Teri Duarte, et al., "Moving Toward Implementation: The Potential for Accountable Care Organizations and Private-Public Partnerships to Advance Active Neighborhood Design," *Preventive Medicine* 69, Supplement 1 (2014): 2.
20. Sacramento County Transportation, *Bikeway Master Plan* (April 2011), https://sacdot.saccounty.net/Pages/BikewayMasterPlan.aspx.
21. City of Sacramento, "Vision Zero," accessed Dec 29, 2020, http://www.cityofsacramento.org/Public-Works/Transportation/Programs-and-Services/Vision-Zero.
22. Sacramento Area Council of Governments, "Sacramento Regional Bike Share," accessed December 29, 2020, https://www.sacog.org/bike-share.
23. M. Alex Hanson, "Measuring the Impact of Complete Streets Projects on Bicyclist and Pedestrian Safety in Sacramento County, CA," Georgia Institute of Technology, School of City and Regional Planning, May 2017, p. 22, 24.
24. Quoted in Thomas Fuller, "A State That Had Very Little Vacancy to Begin with Before Calamity Struck," *New York Times*, November 16, 2018, A10.

5. NEW YORK CITY

1. Tom Farley, *Saving Gotham: A Billionaire Mayor, Activist Doctors, and the Fight for Eight Million Lives* (New York: Norton, 2015), 148, 210, 264, 267, 231, 257, 110, 119.
2. Farley, *Saving Gotham*, 169, 180, 268–69.
3. Justin Davidson, "What New York Can Steal from Hong Kong," *New York Magazine*, October 17, 2011, 106.
4. Thomas R. Frieden, "Take Care New York: A Focused Health Policy," *Journal of Urban Health* 81, no. 3 (2004): 314–16.
5. Thomas R. Frieden, Mary T. Bassett, Lorna E. Thorpe, and Thomas A Farley, "Public Health in New York City, 2002–2007: Confronting

Epidemics of the Modern Era," *International Journal of Epidemiology* 37 (June 2008): 966–97.

6. Farley, *Saving Gotham*, 3, 101.
7. Dan Hirschhorn, "Mayor Steps Up Health Plan," *Wall Street Journal*, July 18, 2013, A17; and Phil Gutis, "Building Healthy Places: A Stairway Renaissance," *Urban Land*, October 17, 2013.
8. Denis Hamill, "Time's Up, Blaz," *New York Daily News*, August 23, 2015, 15; and Linda Stasi, "Bloomy the First Boob," *New York Daily News*, August 23, 2015, 17.
9. Michael M. Grynbaum and Matt Flegenheim, "Mayor de Blasio Raises Prospect of Removing Times Square Pedestrian Plazas," *New York Times*, August 20, 2015.
10. Michael Kimmelman, "Challenging Mayor de Blasio Over Times Square Plazas," *New York Times*, August 21, 2015, C2.
11. See, for example, Jan Gehl, Jeff Risom, and Julia Day, "Times Square: The Naked Truth," *New York Times*, August 31, 2015, A15.
12. Emma G. Fitzsimmons, "City Moves Forward with Restricting Costumed Characters in Times Square," *New York Times*, April 8, 2016, A18; and City of New York, "De Blasio Administration Announces Completion of Times Square Reconstruction" (December 28, 2016), https://www1.nyc.gov/office-of-the-mayor/news/978-16/de-blasio-administration-completion-times-square-reconstruction.
13. Ted Mann, "Street Safety Gains Traction as Political Issue," *Wall Street Journal*, August 7, 2013, A17.
14. Transportation Alternatives, *The Vision Zero Investment: Why New York Must Rebuild Its Most Dangerous Streets Now* (March 19, 2015), 5, https://static1.squarespace.com/static/5cab9d9b65a707a9b36f4b6c/t/609a8fd546baef1013c28b58/1620742114768/The+Vision+Zero+Investment+-+March+2015.pdf.
15. Scott Calvert, "Pedestrian Deaths Highest Since 1990," *Wall Street Journal*, February 28, 2019, A3.
16. Sarah Maslin Nir, "New York City's Pedestrian Deaths Hit Lowest Level Since 1910," *New York Times*, January 8, 2018.
17. Paul Berger, "Opening Up Pedestrian Possibilities," *Wall Street Journal*, April 15, 2019, A10.
18. Stephanos Chen, "Counting Down to a Green New York," *New York Times*, July 14, 2019, 10.

19. Voice of the People, "Readers Sound Off on Mayor Bloomberg and Bicycles and Sen. Ted Cruz," *New York Daily News*, May 5, 2013, 34.
20. Andrew Vesselinovitch, "Pedal Politics," *New York Times*, August 20, 2006.
21. Janette Sadik-Khan and Seth Solomonow, *Street Fight* (New York: Penguin, 2016).
22. Michael M. Grynbaum and Marjorie Connelly, "Bicycle Lanes Draw Wide Support Among New Yorkers, Survey Says," *New York Times*, August 22, 2012, A20.
23. Andrew Tangle, "Bike Lanes Make Cycling in NYC Less Challenging," *Wall Street Journal*, October 9, 2015, H15.
24. Winnie Hu, "More Bicyclists' Deaths Spur a New Safety Plan," *New York Times*, July 25, 2019, A22.
25. Eliot Brown, "The Ride-Hailing Utopia Got Stuck in Traffic," *Wall Street Journal*, February 15–16, 2020, B1, B6.
26. Trevor Boyer, Wes Parnell, and Mikey Light, "Cyclist Carnage Surge," *New York Daily News*, July 2, 2019, 8.
27. Clayton Guse, "Linking Death to Bike Battle," *New York Daily News*, July 21, 2019, 6–7; and Ginia Bellafante, "What Happened to the Bike Safety Push," *New York Times*, July 14, 2019, 29.
28. Winnie Hu and John Surico, "After 3 Cyclist Deaths, Mayor Vows Crackdown," *New York Times*, July 3, 2019, A20; and Hu, "More Bicyclists' Deaths," A22.
29. Jane E. Brody, "Even 'Protected' Bike Lanes Get No Respect," *New York Times*, March 31, 2020, D7.
30. Jon Orcutt, communications director for Bike New York, quoted in Hu, "More Bicyclists' Deaths."
31. Matt Flegenheimer, "For Bloomberg and Bike-Sharing Program, the Big Moment Arrives," *New York Times*, May 27, 2013, A13.
32. Sadik-Khan quoted in Matt A. V. Chaban, "Ex-Transportation Chief is Putting Her 'Streetfight' in Print," *New York Times*, December 6, 2014, A20. See also New York City Department of Transportation, *NYC Bike Share Designed by New Yorkers*, n.d., accessed at https://www.nyc.gov/html/dot/downloads/pdf/bike-share-outreach-report.pdf.
33. Matt Flegenheimer, "Despite Fears, No Riders Were Killed in First 5 Months of Bike-Share Program," *New York Times*, November 5, 2013, A21.

34. CitiBike, http://www.citibikenyc.com/about, accessed January 14, 2016.
35. Citibike NYC website, accessed January 17, 2016, https://www.citibikenyc.com; and "About Citibike," Citibike NYC website, accessed October 18, 2019, https://www.citibikenyc.com/about.
36. Citibike, *September 2019 Monthly Report* (September 2019), p. 3, https://d21xlh2maitm24.cloudfront.net/nyc/September-2019-Citi-Bike-Monthly-Report-marketing.pdf?mtime=20191013094822.
37. Winnie Hu, "Want to Relax in a New York City Park? Join the Crowd," *New York Times*, August 4, 2016, A19.
38. Lisa W. Foderaro, "With 843 Acres Buffed, Central Park Chief Plans to Step Down," *New York Times*, June 7, 2017, A23.
39. *NYC Parks: Framework for an Equitable Future: Program Overview 2014–2015* (2015), p. 28, https://healthyplacesbydesign.org/wp-content/uploads/2015/05/CPI-Program-Summary_Updated.pdf.
40. Tupper Thomas, quoted in Lisa W. Foderaro, "On City Parks, Mayor de Blasio is Seen as Friend, Foe and Something in Between," *New York Times*, June 12, 2015, A19.
41. Benjamin Mueller, "Weekday Car Traffic to Be Banned in Parts of Central and Prospect Parks," *New York Times*, June 19, 2015, A19.
42. NYC Parks website, accessed January 26, 2016, https://www.nycgovparks.org.
43. Greg Hanscom, "Meet the Man Who Re-Energized NYC Parks—and Wants to Do the Same for Yours," *Grist*, October 17, 2014, 3.
44. Foderaro, "On City Parks," A23.
45. C. Summers, L. Cohen, A. Havusha, F. Sliger, and T. Farley, *Take Care New York 2012: A Policy for a Healthier New York City* (New York City Department of Health and Mental Hygiene, September 2009), 13.
46. New York City Obesity Task Force, *Reversing the Epidemic: The New York City Obesity Task Force Plan to Prevent and Control Obesity* (May 31, 2012), 11, http://www.nyc.gov/html/om/pdf/2012/otf_report.pdf.
47. Jennifer H. Cunningham, "Pol Demands Results After Allocating Millions for Greenspaces," *New York Daily News*, September 6, 2013, 35.
48. *NYC Parks: Framework for an Equitable Future.*
49. City of New York, "De Blasio Administration Launches Community Parks Initiative to Build More Inclusive and Equitable Park System"

(October 7, 2014), https://www1.nyc.gov/office-of-the-mayor/news/468-14/de-blasio-administration-launches-community-parks-initiative-build-more-inclusive-equitable#/0.

50. City of New York, "De Blasio Launches Community Parks."
51. City of New York, "Mayor de Blasio Doubles Community Parks Initiative to $285 Million" (October 6, 2015), https://www1.nyc.gov/office-of-the-mayor/news/692-15/mayor-de-blasio-doubles-community-parks-initiative-285-million.
52. Lisa W. Foderaro, "Adding Spaces for Fun and Fitness to a Neglected Park in the South Bronx," *New York Times*, October 19, 2015, A18.
53. Corinne Ramey, "Green Space and Health Linked in Vision for South Bronx," *Wall Street Journal*, September 21, 2015, A15–A16.
54. Lawrence D. Brown and Lawrence R. Jacobs, *The Private Abuse of the Public Interest: Market Myths and Policy Muddles* (Chicago: University of Chicago Press, 2009).
55. Farley, *Saving Gotham*, 37.
56. New York State Senate, "Section 803: Instruction in Physical Education and Kindred Subjects," https://www.nysenate.gov/legislation/laws/EDN/803.
57. New York City Department of Education, "Connect with School Wellness Programs," n.d., accessed December 30, 2020, https://infohub.nyced.org/in-our-schools/programs/health-and-wellness/connect-with-school-wellness-programs.
58. Lillian L. Dunn, Jazmine A. Venturanza, Rhonda J. Walsh, and Cathy A. Nonas, "An Observational Evaluation of Move-To-Improve, a Classroom-Based Physical Activity Program, New York City Schools, 2010," *Preventing Chronic Disease* 9 (2012), http://dx.doi.org/10.5888/pcd9.120072.
59. New York City Parks, "Shape Up NYC," n.d., accessed December 30, 2020, https://www.nycgovparks.org/programs/recreation/shape-up-nyc.
60. "Active Design Tool Kit—DRAFT," n.d., www.activedesigntoolkit.wordpress.com/policy-program-examples/education-and-school-construction-departments/.
61. Charles Di Maggio and Guohua Li, "Effectiveness of a Safe Routes to School Program in Preventing School-Age Pedestrian Injury," *Pediatrics* 131 (2013): 290–96.

62. Devin Madden, one-page handout, "Active Design Guidelines for Schools," n.d. See also The Partnership for a Healthier New York City, *Active Design Toolkit for Schools* (2015), https://centerforactivedesign.org/dl/schools.pdf.
63. Scott M. Stringer, "Dropping the Ball: Disparities in Physical Education in New York City's Schools," New York City Comptroller, May 2015, https://comptroller.nyc.gov/reports/dropping-the-ball-disparities-in-physical-education-in-new-york-city/.
64. Katie Honan, "Mayor Keeps His Focus on Economic Gap," *Wall Street Journal*, January 11, 2019, A9A.
65. See Kimberley Roussin Isett, Miriam J. Laugesen, and David H. Cloud, "Learning from New York City: A Case Study of Public Health Policy Practice in the Bloomberg Administration," *Journal of Public Health Management Practice* 21, no. 4 (2015): 313–22.

6. EVALUATION MEETS IMPLEMENTATION

The subtitle of the chapter, "the struggle for the real," comes from Clifford Geertz, *Islam Observed: Religious Development in Morocco and Indonesia* (Chicago: University of Chicago Press, 1968), chap. 4.

1. MacDonald, *Changing Places*, 12.
2. Cornelia Guell, Roger Mackett, and David Ogilvie, "Negotiating Multisectoral Evidence: A Qualitative Study of Knowledge Exchange at the Intersection of Transport and Public Health," *BMC Public Health* 17 (2017): 10, 8.
3. Ditte Heering Holt, Morten Hulvej Rod, Susanne Boch Waldorff, and Tine Tjornhoj-Thomsen, "Elusive Implementation: An Ethnographic Study of Intersectoral Policymaking for Health," *BMC Health Services Research* 18 (2018): 2, 1.
4. Jerome Bruner, *Acts of Meaning* (Cambridge, Mass.: Harvard University Press, 1990), 108, xiii.
5. Jerry Z. Muller, *The Tyranny of Metrics* (Princeton, N.J.: Princeton University Press, 2018), 134–35.
6. Richard Elmore and Milbrey Wallin McLaughlin, "The Federal Role in Education: Learning from Experience," *Education and Urban Society* 15 (May 1983): 315.

7. Robert Wood Johnson Foundation, *Active Living by Design*, 15.
8. David Stockman, "The Social Pork Barrel," *The Public Interest* 39 (Spring 1975).
9. On methodolatry, see Bruner, *Acts of Meaning*, xi. On sense-making, see James C. Scott, *Two Cheers for Anarchism* (Princeton, N.J.: Princeton University Press, 2012), 147. See also Muller, *The Tyranny of Metrics*; and Henry Mintzberg, *Managing the Myths of Health Care*. Oakland, Calif.: Bennett-Koehler, 2017.
10. Bruner, *Acts of Meaning*, 60, 30.
11. Theodore J. Lowi, "Four Systems of Policy, Politics, and Choice," *Public Administration Review* 32, no. 4 (1972): 298–310; and E. E. Schattschneider, *Politics, Pressures and the Tariff: A Study of Free Private Enterprise in Pressure Politics, As Shown in the 1929–1930 Revision of the Tariff* (New York: Prentice Hall, 1935).
12. See, e.g., Theodore J. Lowi, "American Business, Public Policy, Case Studies, and Political Theory," *World Politics* 16, no. 4 (1964): 677–715; Lowi, "Four Systems"; and James Q. Wilson, *Political Organizations* (New York: Basic Books, 1973).
13. John E. Wennberg, *Tracking Medicine: A Researcher's Quest to Understand Health Care* (New York: Oxford University Press, 2010).
14. Kimberly J. Morgan and Andrea Louise Campbell, *The Delegated Welfare State: Medicare, Markets and the Governance of Social Policy* (New York: Oxford University Press, 2011); and Jamila Michener, *Fragmented Democracy: Medicaid, Federalism, and Unequal Politics* (New York: Cambridge University Press, 2018).
15. Rykwert, *The Seduction of Place*, 246.
16. William B. Neenan, *Political Economy of Urban Areas* (Chicago: Markham, 1972); and Lewis, *Divided Highways*, 80.
17. On the "culture of fear," see Barry Glassner, *The Culture of Fear: Why Americans Are Afraid of the Wrong Things* (New York: Basic Books, 1999).
18. Harry Eyres, *Horace and Me: Life Lessons from an Ancient Poet* (New York: Farrar, Straus and Giroux, 2013), 151.
19. Bruner, *Acts of Meaning*, 95.
20. James Agee, *Cotton Tenants: Three Families* (Brooklyn: Melville House, 2014), 220.
21. Jill S. Litt, Hannah L. Reed, Rachel G. Tabak, Susan G. Zieff, Amy A. Eyler, Rodney Lyn, Karin Valentine Goins, Jeanette Gustat,

and Nancy O'Hara Tompkins, "Active Living Collaborative in the United States: Understanding Characteristics, Activities, and Achievement of Environmental and Policy Change," *Preventing Chronic Disease* 10 (2013), https://doi.org/10.5888/pcd10.120162.

22. On the lack of "conceptual tools to assess and compare systematically the barriers to effective vertical coordination" that confront the implementers in a range of policies, see Christian Adam, Steffen Hurka, Christoph Knill, B. Guy Peters, and Yves Steinebach, "Introducing Vertical Policy Coordination to Comparative Policy Analysis: The Missing Link Between Policy Production and Implementation," *Journal of Comparative Policy Analysis* 21, no. 5 (2019): 499.
23. Vanderbilt, *Traffic*, 112.
24. Lewis, *Divided Highways*, 241–43.
25. Altshuler and Luberoff, *Mega-Projects*, 121.
26. Sacramento Area Council of Governments, "Sacramento Region Blueprint," accessed December 28, 2020, https://www.sacog.org/sacramento-region-blueprint.
27. On the problem—and promise—of "environmental imageability" and its connection to "purposeful mobility," see Kevin Lynch, *The Image of the City* (Cambridge, Mass.: MIT Press, 1960), 13, 124.
28. For an instructive case in point, see Mick Cornett and Jayson White, *The Next American City: The Big Promise of Our Midsize Metros* (New York: Putnam's, 2018).
29. John Kingdon, *Agendas, Alternatives, and Public Policy*, 2nd ed. (London: Longman, 2011).
30. Clarence N. Stone, *Regime Politics: Governing Atlanta 1946–1988* (Lawrence: University Press of Kansas, 1989).
31. Simon Kuper, "Paris in 2050—from Great City to New Metropolis," *Financial Times*, March 14–15, 2020, 19.
32. Ralph Buehler, John Pucher, Regine Gerike, and Thomas Götschi, "Reducing Car Dependence in the Heart of Europe: Lessons from Germany, Austria, and Switzerland," *Transport Reviews* 37 (2017): 24.
33. Buehler et al., "Reducing Car Dependence," 13–18.
34. Buehler et al., "Reducing Car Dependence," 15.
35. Ralph Buehler, John Pucher, and Alan Altshuler, "Vienna's Path to Sustainable Transport," *International Journal of Sustainable Transportation* 11 (2017): 268.

36. Buehler et al., "Vienna's Path," 261, 268–69.
37. Paul Collier, *The Future of Capitalism: Facing the New Anxieties* (New York: HarperCollins, 2018), 3, and 125–53.
38. Geertz, *Islam Observed*, 7.
39. Jefferson Cowie, *The Great Exception: The New Deal and the Limits of American Politics* (Princeton, N.J.: Princeton University Press, 2016), 229.

CONCLUSION

1. James C. Scott, *Seeing Like a State: How Certain Schemes to Improve the Human Condition Have Failed* (New Haven, Conn.: Yale University Press, 1998).
2. James Q. Wilson, *Bureaucracy: What Government Agencies Do and Why They Do It* (New York: Basic Books, 1989): 99–100; and Martha Derthick, *Policymaking for Social Security* (Washington, DC: Brookings Institution, 1979).
3. Deborah Stone, *The Disabled State* (Philadelphia: Temple University Press, 1984); and Lael R. Keiser, "Street-Level Bureaucrats' Decision-Making: Determining Eligibility in the Social Security Disability Program," *Public Administration Review* 70 (2010): 247–57. On "near chaos," see Wilson, *Bureaucracy*, 100.
4. Robert Pear, "Collecting Disability? Uncle Sam May Not Be a Friend on Social Media," *New York Times*, March 11, 2019, A13.
5. Miriam J. Laugesen, *Fixing Medical Prices: How Physicians Are Paid* (Cambridge, Mass.: Harvard University Press, 2016).
6. Michael Sparer, *Medicaid and the Limits of State Health Reform* (Philadelphia: Temple University Press, 1996); Michael S. Sparer, "Federalism and the Patient Protection and Affordable Care Act of 2010: The Founding Fathers Would Not Be Surprised," *Journal of Health Politics, Policy and Law* 36 (June 2011) : 461–68; and David G. Smith and Judith D. Moore, *Medicaid Politics and Policy: 1965–2007* (Piscataway, N.J.: Transaction, 2009).
7. Michael Lipsky, *Street-Level Bureaucracy: Dilemmas of the Individual in Public Services* (New York: Russell Sage Foundation, 2010).
8. William J. Baumol, *The Cost Disease: Why Computers Get Cheaper and Health Care Doesn't* (New Haven, Conn.: Yale University Press, 2012).

9. Daniel L. Hatcher, *The Poverty Industry: The Exploitation of America's Most Vulnerable Citizens* (New York: New York University Press, 2016), 25.
10. Scott, *Seeing Like a State*.
11. Burke, *On Symbols and Society*, 59.
12. Esther Duflo, "Richard T. Ely Lecture: The Economist as Plumber," *American Economic Review* 107, no. 5, (2017).
13. Peter deLeon and Linda deLeon, "Whatever Happened to Policy Implementation? An Alternative Approach," *Journal of Public Administration Research and Theory* 12 (October 2002): 467.
14. deLeon and deLeon, "Whatever Happened," 468–72.
15. deLeon and deLeon, "Whatever Happened," 467.
16. Peter L. Hupe and Harald Saetren, "The Sustainable Future of Implementation Research: On the Development of the Field and its Paradoxes," *Public Policy and Administration* 29 (2014): 77.
17. Hupe and Saetren, "The Sustainable Future," 84, 86.
18. Stephen Roll, Stephanie Moulton, and Jodi Sandfort, "A Comparative Analysis of Two Streams of Implementation Research," *Journal of Public and Nonprofit Affairs* 3 (2017): 16, 12, 5.
19. deLeon and deLeon, "Whatever Happened," 489.
20. Richard E. Matland, "Synthesizing the Implementation Literature: The Ambiguity-Conflict Model of Policy Implementation," *Journal of Public Administration* 5 (1995): 146, 170; and Harald Saetren, "Facts and Myths About Research on Public Policy Implementation: Out-of-Fashion, Allegedly Dead, But Still Very Much Alive and Relevant," *Policy Studies Journal* 33 (2005): 562, 573.
21. Lucie Cerna, *The Nature of Policy Change and Implementation: A Review of Different Theoretical Approaches* (Paris: OECD, 2013), 22–23.
22. Robert Merton, "Bureaucratic Structure and Personality," *Social Forces* 17 (1940): 560–68.
23. Pressman and Wildavsky, *Implementation*.
24. Jerome Murphy, "Title I of ESEA: The Politics of Implementing Federal Education Reform," *Harvard Educational Review* 41 (April 1971): 35–63.
25. Daniel Patrick Moynihan, *Maximum Feasible Misunderstanding: Community Action in the War on Poverty* (New York: Free Press, 1969).

26. Malcolm L. Goggin, *Policy Design and the Politics of Implementation: The Case of Child Health Care in the American States*. Knoxville: University of Tennessee Press, 1987.
27. Peter M. Blau, *The Dynamics of Bureaucracy: A Study of Interpersonal Relationships in Two Government Agencies* rev. ed. (Chicago: University of Chicago Press, 1963), 55–56, 29.
28. Richard P. Nathan, *Turning Promises into Performance: The Management Challenges of Implementing Workfare* (New York: Columbia University Press, 1993), 126.
29. Milbrey Wallin McLaughlin, "Learning from Experience: Lessons from Policy Implementation," *Educational Evaluation and Policy Analysis* 9 (1987): 172.
30. Theodore J. Lowi, *The End of Liberalism: Ideology, Policy, and the Crisis of Public Authority* (New York: Norton, 1969).
31. In a similar vein, see Matland, "Synthesizing the Implementation Literature," 147–48, 155, 158–59.
32. Elmore and McLaughlin, "Federal Role in Education," 313.
33. Theda Skocpol, *Social Policy in the United States: Future Possibilities in Historical Perspective* (Princeton, N.J.: Princeton University Press, 1995), 250–74; and Andrea Louise Campbell, *Trapped in America's Safety Net: One Family's Struggle* (Chicago: University of Chicago Press, 2014).
34. Campbell, *Trapped in America's Safety Net*, 69.
35. Elizabeth Davidson, Randall Reback, Jonah Rockoff, and Heather L. Schwartz, "Fifty Ways to Leave a Child Behind: Idiosyncrasies and Discrepancies in States' Implementation of NCLB," *Educational Researcher* 44 (2015): 347, 356.
36. Michener, *Fragmented Democracy*, 34.
37. Morgan and Campbell, *The Delegated Welfare State.*
38. Michael Lewis, *The Fifth Risk* (New York: Norton, 2018), 100.
39. Adam Oliver, *The Origins of Behavioral Public Policy* (Cambridge: Cambridge University Press, 2017), 103, 157–58, 173.
40. Esther Duflo and Abhijit Banerjee, "Economic Incentives Don't Always Do What We Want Them To," *New York Times*, October 26, 2019.
41. On the problem of "disembodied incentives," see Lawrence D. Brown *Politics and Health Care Organization: HMOs as Federal Policy* (Washington, D.C.: Brookings Institution, 1983).

42. Philip Selznick, *Leadership in Administration* (Evanston, Ill.: Row, Peterson, 1957).
43. Daniel Simonet, "The New Public Management Theory in the British Health Care System: A Critical Review," *Administration and Society* 47 (2015): 802–26.
44. Alain Enthoven, *Health Plan: The Only Practical Solution to the Soaring Costs of Medical Care* (Reading, Mass.: Addison-Wesley, 1980); and Alain Enthoven, *Theory and Practice of Managed Competition* (Amsterdam: Elsevier, 1988).
45. Jonathan Hammond, Colin Lorne, Anna Coleman, Pauline Allen, Nicholas Mays, Rinita Dam, Thomas Mason, and Kath Checkland, "The Spatial Politics of Place and Health Policy: Exploring Sustainability and Transformation Plans in the English NHS," *Social Science and Medicine* 190 (October 2017): 12.
46. For a deeply insightful investigation of early competitive ventures, see Michael I. Harrison, *Implementing Change in Health Systems: Market Reforms in the United Kingdom, Sweden, and the Netherlands* (London: Sage, 2004).
47. Hans Maarse, Patrick Jeurissen, and Dirk Ruwaard, "Results of the Market-Oriented Reform in the Netherlands: A Review," *Health Economics, Policy, and Law* 11 (April 2016): 167–78.
48. Brown and Jacobs, *Private Abuse of Public Interest*, 38–66.
49. Peter L. Hupe and Michael J. Hill, "'And the Rest Is Implementation': Comparing Approaches to What Happens in Policy Processes Beyond Great Expectations," *Public Policy and Administration* 31 (2016): 103.
50. Peter L. Hupe, "The Thesis of Incongruent Implementation: Revisiting Pressman and Wildavsky," *Public Policy and Administration* 26, (2011): 64.
51. Hupe, "The Thesis of Incongruent Implementation," 67.
52. Hupe, "The Thesis of Incongruent Implementation," 66.
53. Hupe, "The Thesis of Incongruent Implementation," 69, 71.
54. Hupe, "The Thesis of Incongruent Implementation," 73.
55. Hupe, "The Thesis of Incongruent Implementation," 71.
56. Hupe, "The Thesis of Incongruent Implementation," 72.
57. Hupe, "The Thesis of Incongruent Implementation," 73, 76. For an artful implementation study that highlights capacities and attitudes within federal agencies, see Steven Kelman, "Using Implementation

Research to Solve Implementation Problems: The Case of Emergency Energy Assistance," *Journal of Policy Analysis and Management* 4 (Fall 1984): 75–91.

58. The phrase "an interwoven somewhat," is taken from Marianne Moore, *The Monkey Puzzler, in Observations* (New York: Farrar, Straus and Giroux, 2016), 25.
59. Hupe, "The Thesis of Incongruent Implementation," 76.
60. Quoted in deLeon and deLeon, "Whatever Happened," 473.
61. Michel Foucault, *Society Must Be Defended* (New York: Picador, 2003), 32.
62. James P. Spillane, Brian J. Reiser, and Todd Reimer, "Policy Implementation and Cognition: Reframing and Refocusing Implementation Research," *Review of Educational Research* 72 (2002): 387–431; and Michael D. Siciliano, Nienke M. Moolenaar, Alan J. Daly, and Yi-Hwa Liou, "A Cognitive Perspective on Policy Implementation: Reform Beliefs, Sensemaking, and Social Networks," *Public Administration Review* 77 (2017): 889–901.
63. On "sensitizing concepts," see Herbert Blumer, "What Is Wrong with Social Theory," *American Sociological Review* 18 (1954): 3–10; on "metis," see Scott, *Seeing Like a State*, chap. 9; on "sense-making," see Karl E. Weick, *Sensemaking in Organizations* (Thousand Oaks, Calif.: Sage, 1995); and on "thick description," see Geertz, *Interpretation of Cultures.*
64. Guell et al., "Negotiating Multisectoral Evidence," 6.
65. Geertz, *Life Among the Anthros*, 239.
66. Susan Watt, Wendy Sword, and Paul Krueger. "Implementation of a Health Care Policy: An Analysis of Barriers and Facilitators to Practice Change." *BMC Health Services Research* 5 (2005): 8.
67. Trisha Greenhalgh, "What Is This Knowledge that We Seek to 'Exchange?'" *Milbank Quarterly* 88 (2010): 496.
68. Stone, "Policy Paradox," 13.
69. Matthew Stewart, *The Management Myth: Debunking Modern Business Philosophy* (New York: Norton, 2009).
70. Joan Didion, *Democracy* (New York: Simon and Schuster, 1984), 215.
71. Alain Pessin, *The Sociology of Howard Becker: Theory with a Wide Horizon* (Chicago: University of Chicago Press, 2017), 122.
72. Stone, "Policy Paradox," 28.

73. Pessin, *Sociology of Howard Becker*, xi. See also Hugh Heclo, *Modern Social Politics in Britain and Sweden* (New Haven, Conn.: Yale University Press, 1974), chap. 6; and Rudolf Klein, *The New Politics of the NHS* (Oxford: Radcliffe Publishing, 2010), 285.
74. Hendrik Wagenaar, "'Knowing' the Rules: Administrative Work as Practice," *Public Administration Review* 64 (2004): 643.
75. Wagenaar, "'Knowing' the Rules," 644.
76. Pessin, *Sociology of Howard Becker*, 44.
77. Pessin, *Sociology of Howard Becker*, 83, 111.
78. Pessin, *Sociology of Howard Becker*, 60.
79. Stewart, *The Management Myth*, 210.
80. Eleanor Farrar, John E. DeSanctis, and David K. Cohen, "The Lawn Party: The Evolution of Federal Programs in Local Settings," *Phi Delta Kappan* 62 (1980): 170–71.
81. Spillane et al., "Policy Implementation and Cognition," 413, 394.
82. A. J. Liebling, *Between Meals: An Appetite for Paris* (New York: Farrar, Straus and Giroux, 1986), 22.

REFERENCES

Abbott, Carl. "Five Downtown Strategies: Policy Discourses and Downtown Planning Since 1945." In *Urban Public Policy: Historical Modes and Methods*, ed. Martin V. Melosi, 5–27. University Park: Pennsylvania State Press, 1993.

Adam, Christian, Steffen Hurka, Christoph Knill, B. Guy Peters, and Yves Steinebach. "Introducing Vertical Policy Coordination to Comparative Policy Analysis: The Missing Link Between Policy Production and Implementation." *Journal of Comparative Policy Analysis* 21, no. 5 (2019): 499–517.

Agee, James. *Cotton Tenants: Three Families*. Brooklyn: Melville House, 2014.

Altshuler, Alan and David Luberoff. *Mega-Projects: The Changing Politics of Urban Public Investment*. Washington, D.C.: Brookings Institution, 2003.

Bailey, Philip M. "Fischer Unveils $1.4B Transportation Plan." *Louisville Courier-Journal*, April 14, 2016. https://www.courier-journal.com/story/news/politics/metro-government/2016/04/04/plan-citys-mobility-future-must-car-lite/82998494.

Bassett, David R., Jr., Holly R. Wyatt, Helen Thompson, John C. Peters, and James O. Hill. "Pedometer-Measured Physical Activity and Health Behaviors in United States Adults." *Medicine and Science in Sports and Exercise* 42, no. 19 (2010): 1819–25.

Baumol, William J. *The Cost Disease: Why Computers Get Cheaper and Health Care Doesn't*. New Haven, Conn.: Yale University Press, 2012.

Bellafante, Ginia. "What Happened to the Bike Safety Push." *New York Times*, July 14, 2019, 29.

Berger, Paul. "Opening Up Pedestrian Possibilities." *Wall Street Journal*, April 15, 2019, A10.

Berryman, Jack W. "Motion and Rest: Galen on Exercise and Health." *Lancet* 380, no. 9838 (2012): 210–11.

Blau, Peter M. *The Dynamics of Bureaucracy: A Study of Interpersonal Relationships in Two Government Agencies*, rev. ed. Chicago: University of Chicago Press, 1963.

Blumer, Herbert. "What Is Wrong with Social Theory." *American Sociological Review* 18 (1954): 3–10.

Boyer, Trevor, Wes Parnell, and Mikey Light. "Cyclist Carnage Surge." *New York Daily News*, July 2, 2019, 8.

Brasuell, James. "Coming This Spring: Bikeshare Pilot Program in Albuquerque." *Planetizen*, February 28, 2015. https://www.planetizen.com/node/74567.

Brody, Jane E. "Even 'Protected' Bike Lanes Get No Respect." *New York Times*, March 31, 2020, D7.

Brown, Alicia, Project Coordinator, Walks Sacramento. Letter to Garrett Norman, Assistant Planner, City of Sacramento Community Development Department (December 8, 2017). https://www.walksacramento.org/wp-content/uploads/2018/01/Spanos-Natomas-Crossing-WALKSac-20171208-Letter.pdf.

Brown, Eliot. "The Ride-Hailing Utopia Got Stuck in Traffic." *Wall Street Journal*, February 15–16, 2020, B1, B6.

Brown, Lawrence D. *Politics and Health Care Organization: HMOs as Federal Policy*. Washington, D.C.: Brookings Institution, 1983.

Brown, Lawrence D., and Lawrence R. Jacobs. *The Private Abuse of the Public Interest: Market Myths and Policy Muddles*. Chicago: University of Chicago Press, 2009.

Bruner, Jerome. *Acts of Meaning*. Cambridge, Mass.: Harvard University Press, 1990.

Buehler, Ralph, and John Pucher. "Urban Transport: Promoting Sustainability in Germany." In *Lessons from Europe? What Americans Can Learn from European Public Policies*, ed. R. Daniel Keleman. Los Angeles: Sage, 2014.

Buehler, Ralph, John Pucher, and Alan Altshuler. "Vienna's Path to Sustainable Transport." *International Journal of Sustainable Transportation* 11 (2017): 257–71.

Buehler, Ralph, John Pucher, Regine Gerike, and Thomas Götschi. "Reducing Car Dependence in the Heart of Europe: Lessons from Germany, Austria, and Switzerland." *Transport Reviews* 37 (2017): 4–28.

Burke, Kenneth. *On Symbols and Society*, ed. Joseph Gusfield. Chicago: University of Chicago Press, 1989.

Calvert, Scott. "Pedestrian Deaths Highest Since 1990." *Wall Street Journal*, February 28, 2019, A3.

Campbell, Andrea Louise. *Trapped in America's Safety Net: One Family's Struggle.* Chicago: University of Chicago Press, 2014.

Campos, Paul. *The Obesity Myth: Why America's Obsession with Weight Is Dangerous to Your Health.* New York: Gotham, 2004.

Cardinal, Bradley J., and Kris Day. "Special Issue: Active Living Research." *American Journal of Health Promotion* 21 (2007): Supplement.

Centers for Disease Control and Prevention and the American College of Sports Medicine. "A Recommendation." *Journal of the American Medical Association* 273 (1995): 402–7.

Cerna, Lucie. *The Nature of Policy Change and Implementation: A Review of Different Theoretical Approaches.* Paris: OECD, 2013.

Chaban, Matt A. V. "Ex-Transportation Chief Is Putting Her 'Streetfight' in Print." *New York Times*, December 6, 2014, A20.

Chen, Stephanos. "Counting Down to a Green New York." *New York Times*, July 14, 2019, 10.

Chokshi, Niraj. "Population Growth in New Mexico Is Approaching Zero." *Washington Post*, January 17, 2014.

CitiBike. *September 2019 Monthly Report.* September 2019. https://d21xlh2maitm24.cloudfront.net/nyc/September-2019-Citi-Bike-Monthly-Report-marketing.pdf?mtime=20191013094822.

City of Albuquerque. "Integrated Development Ordinance." 2017, as amended. http://www.cabq.gov/planning/codes-policies-regulations/integrated-development-ordinance.

——. "Prescription Trails." N.d. http://www.cabq.gov/parks/prescription-trails.

City of New York. "De Blasio Administration Announces Completion of Times Square Reconstruction." December 28, 2016. https://www1.nyc.gov/office-of-the-mayor/news/978-16/de-blasio-administration-completion-times-square-reconstruction.

——. "De Blasio Administration Launches Community Parks Initiative to Build More Inclusive and Equitable Park System." October 7, 2014. https://www1.nyc.gov/office-of-the-mayor/news/468-14/de-blasio-administration-launches-community-parks-initiative-build-more-inclusive-equitable#/0.

——. "Mayor de Blasio Doubles Community Parks Initiative to $285 Million." October 6, 2015. https://www1.nyc.gov/office-of-the-mayor/news/692-15/mayor-de-blasio-doubles-community-parks-initiative-285-million.

Collier, Paul. *The Future of Capitalism: Facing the New Anxieties*. New York: HarperCollins, 2018.

Complete Streets Ordinance. F/S O-14-27, City of Albuquerque Twenty-First Council (2015). https://www.cabq.gov/council/documents/councilor-district-2-documents/CompleteStreetsLegislationPacket.pdf.

Cornett, Mick, and Jayson White. *The Next American City: The Big Promise of Our Midsize Metros*. New York: Putnam's, 2018.

Cowie, Jefferson. *The Great Exception: The New Deal and the Limits of American Politics*. Princeton, N.J.: Princeton University Press, 2016.

Cunningham, Jennifer H. "Pol Demands Results After Allocating Millions for Greenspaces." *New York Daily News*, September 6, 2013, 35.

Davidson, Elizabeth, Randall Reback, Jonah Rockoff, and Heather L. Schwartz. "Fifty Ways to Leave a Child Behind: Idiosyncrasies and Discrepancies in States' Implementation of NCLB." *Educational Researcher* 44 (2015): 347–58.

Davidson, Justin. "What New York Can Steal from Hong Kong." *New York Magazine*, October 17, 2011, 106.

deLeon, Peter, and Linda deLeon. "Whatever Happened to Policy Implementation? An Alternative Approach." *Journal of Public Administration Research and Theory* 12 (October 2002): 467–92.

Derthick, Martha. *Policymaking for Social Security*. Washington, D.C.: Brookings Institution, 1979.

Didion, Joan. *Democracy*. New York: Simon and Schuster, 1984.

Dille, Ian. "The 50 Best Bike Cities of 2016." *Bicycling*, September 19, 2016. www.bicycling.com/culture/news/the-50-best-bike-cities-of-2016.

Di Maggio, Charles, and Guohua Li. "Effectiveness of a Safe Routes to School Program in Preventing School-Age Pedestrian Injury." *Pediatrics* 131 (2013): 290–96.

Dolinsky, Joe. "Public Square Business Owners Look for Foot Traffic." *Wilkes-Barre Times Leader*, June 14, 2015, 2A.

Duflo, Esther. "Richard T. Ely Lecture: The Economist as Plumber." *American Economic Review* 107, no. 5 (2017): 1–26.

Duflo, Esther, and Abhijit Banerjee. "Economic Incentives Don't Always Do What We Want Them To." *New York Times*, October 26, 2019.

Dunn, Lillian L., Jazmine A. Venturanza, Rhonda J. Walsh, and Cathy A. Nonas. "An Observational Evaluation of Move-To-Improve, a Classroom-Based Physical Activity Program, New York City Schools, 2010." *Preventing Chronic Disease* 9 (2012). http://dx.doi.org/10.5888/pcd9.120072.

Elmore, Richard, and Milbrey Wallin McLaughlin. "The Federal Role in Education: Learning from Experience." *Education and Urban Society* 15 (May 1983): 309–30.

Enrenhalt, Alan. *The Lost City: The Forgotten Virtues of Community in America.* New York: Basic Books, 1995.

Enthoven, Alain. *Health Plan: The Only Practical Solution to the Soaring Costs of Medical Care.* Reading, Mass.: Addison-Wesley, 1980.

——. *Theory and Practice of Managed Competition.* Amsterdam: Elsevier, 1988.

Eyres, Harry. *Horace and Me: Life Lessons from an Ancient Poet.* New York: Farrar, Straus and Giroux, 2013.

Farley, Tom. *Saving Gotham: A Billionaire Mayor, Activist Doctors, and the Fight for Eight Million Lives.* New York: Norton, 2015.

Farrar, Eleanor, John E. DeSanctis, and David K. Cohen. "The Lawn Party: The Evolution of Federal Programs in Local Settings." *Phi Delta Kappan* 62 (1980): 167–71.

Fitzsimmons, Emma G. "City Moves Forward with Restricting Costumed Characters in Times Square." *New York Times*, April 8, 2016, A18.

Flegenheimer, Matt. "Despite Fears, No Riders Were Killed in First 5 Months of Bike-Share Program." *New York Times*, November 5, 2013, A21.

——. "For Bloomberg and Bike-Sharing Program, the Big Moment Arrives." *New York Times*, May 27, 2013, A13.

Foderaro, Lisa W. "Adding Spaces for Fun and Fitness to a Neglected Park in the South Bronx." *New York Times*, October 19, 2015, A18.

——. "On City Parks, Mayor de Blasio is Seen as Friend, Foe and Something in Between." *New York Times*, June 12, 2015, A19.

——. "With 843 Acres Buffed, Central Park Chief Plans to Step Down." *New York Times*, June 7, 2017, A23.

Foucault, Michel. *Society Must Be Defended*. New York: Picador, 2003.

Frieden, Thomas R. "Take Care New York: A Focused Health Policy." *Journal of Urban Health* 81, no. 3 (2004): 314–16.

Frieden, Thomas R., Mary T. Bassett, Lorna E. Thorpe, and Thomas A Farley. "Public Health in New York City, 2002–2007: Confronting Epidemics of the Modern Era." *International Journal of Epidemiology* 37 (June 2008): 966–97.

Fuller, Thomas. "A State That Had Very Little Vacancy to Begin with Before Calamity Struck." *New York Times*, November 16, 2018, A10.

Gaesser, Glenn. *Big Fat Lies: The Truth About Your Weight and Your Health*. Carlsbad, Calif.: Gurze, 2002.

Gans, Herbert J. *The Urban Villagers: Group and Class in the Life of Italian-Americans*. New York: Free Press, 1962.

Gay, Peter. *Weimar Culture: The Outsider as Insider*. New York: Norton, 2001.

Geertz, Clifford. *The Interpretation of Cultures*. New York: Basic Books, 1973.

——. *Islam Observed: Religious Development in Morocco and Indonesia*. Chicago: Chicago University Press, 1968.

——. *Life Among the Anthros and Other Essays*, ed. Fred Inglis. Princeton, N.J.: Princeton University Press, 2010.

Gehl, Jan, Jeff Risom, and Julia Day. "Times Square: The Naked Truth." *New York Times*, August 31, 2015, A15.

Gilman, Sander. *Franz Kafka: The Jewish Patient*. New York: Routledge, 1995.

Glassner, Barry. *The Culture of Fear: Why Americans Are Afraid of the Wrong Things*. New York: Basic Books, 1999.

Goggin, Malcolm L. *Policy Design and the Politics of Implementation: The Case of Child Health Care in the American States*. Knoxville: University of Tennessee Press, 1987.

Greenhalgh, Trisha. "What Is This Knowledge That We Seek to 'Exchange?'" *Milbank Quarterly* 88 (2010): 492–99.

Grynbaum, Michael M., and Marjorie Connelly. "Bicycle Lanes Draw Wide Support Among New Yorkers, Survey Says." *New York Times*, August 22, 2012, A20.

Grynbaum, Michael M., and Matt Flegenheim. "Mayor de Blasio Raises Prospect of Removing Times Square Pedestrian Plazas." *New York Times*, August 20, 2015.

Guell, Cornelia, Roger Mackett, and David Ogilvie. "Negotiating Multisectoral Evidence: A Qualitative Study of Knowledge Exchange at the Intersection of Transport and Public Health." *BMC Public Health* 17 (2017).

Guse, Clayton. "Linking Death to Bike Battle." *New York Daily News*, July 21, 2019, 6–7.

Gutis, Phil. "Building Healthy Places: A Stairway Renaissance." *Urban Land*, October 17, 2013.

Hamill, Denis. "Time's Up, Blaz." *New York Daily News*, August 23, 2015, 15.

Hammond, Jonathan, Colin Lorne, Anna Coleman, Pauline Allen, Nicholas Mays, Rinita Dam, Thomas Mason, and Kath Checkland. "The Spatial Politics of Place and Health Policy: Exploring Sustainability and Transformation Plans in the English NHS." *Social Science and Medicine* 190 (October 2017): 217–26.

Hanscom, Greg. "Meet the Man Who Re-Energized NYC Parks—and Wants to Do the Same for Yours." *Grist*, October 17, 2014, 3.

Hanson, M. Alex. "Measuring the Impact of Complete Streets Projects on Bicyclist and Pedestrian Safety in Sacramento County, CA." Georgia Institute of Technology, School of City and Regional Planning, May 2017, 1–30.

Harrison, Michael I. *Implementing Change in Health Systems: Market Reforms in the United Kingdom, Sweden, and the Netherlands*. London: Sage, 2004.

Hatcher, Daniel L. *The Poverty Industry: The Exploitation of America's Most Vulnerable Citizens*. New York: New York University Press, 2016.

Heclo, Hugh. *Modern Social Politics in Britain and Sweden*. New Haven, Conn.: Yale University Press, 1974.

Hett, Benjamin Carter. *The Death of Democracy: Hitler's Rise to Power and the Downfall of the Weimar Republic*. New York: St. Martin's Griffin, 2018.

Hirschhorn, Dan. "Mayor Steps Up Health Plan," *Wall Street Journal*, July 18, 2013, A17.

Holt, Ditte Heering, Morten Hulvej Rod, Susanne Boch Waldorff, and Tine Tjornhoj-Thomsen. "Elusive Implementation: An Ethnographic

Study of Intersectoral Policymaking for Health." *BMC Health Services Research* 18 (2018).

Honan, Katie. "Mayor Keeps His Focus on Economic Gap." *Wall Street Journal*, January 11, 2019, A9A.

Hu, Winnie. "More Bicyclists' Deaths Spur a New Safety Plan." *New York Times*, July 25, 2019, A22.

——. "Want to Relax in a New York City Park? Join the Crowd." *New York Times*, August 4, 2016, A19.

Hu, Winnie, and John Surico. "After 3 Cyclist Deaths, Mayor Vows Crackdown." *New York Times*, July 3, 2019, A20.

Huizinga, Johan. *Homo Ludens: A Study of the Play Element in Culture*. Boston: Beacon, 1950.

Hupe, Peter L. "The Thesis of Incongruent Implementation: Revisiting Pressman and Wildavsky." *Public Policy and Administration* 26 (2011): 63–80.

Hupe, Peter L., and Michael J. Hill. "'And the Rest Is Implementation': Comparing Approaches to What Happens in Policy Processes Beyond Great Expectations." *Public Policy and Administration* 31 (2016): 103–21.

Hupe, Peter L., and Harald Saetren. "The Sustainable Future of Implementation Research: On the Development of the Field and its Paradoxes." *Public Policy and Administration* 29 (2014): 77–83.

Isett, Kimberley Roussin, Miriam J. Laugesen, and David H. Cloud. "Learning from New York City: A Case Study of Public Health Policy Practice in the Bloomberg Administration." *Journal of Public Health Management Practice* 21, no. 4 (2015): 313–22.

Jacobs, Jane. *The Death and Life of Great American Cities*. New York: Random House, 1961.

Joint Urban Studies Center. *Healthy Communities: Luzerne County Assessment*. Wilkes-Barre, Penn.: Joint Urban Studies Center, May 25, 2007.

Jones, Sherry Everett, and Sliwa, Sarah. "School Factors Associated with the Percentage of Students Who Walk or Bike to School, School Health Policies and Practices Study 2014." *Preventing Chronic Disease 13* (2016): 1–8.

Keiser, Lael R. "Street-Level Bureaucrats' Decision-Making: Determining Eligibility in the Social Security Disability Program." *Public Administration Review* 70 (2010): 247–57.

Kelman, Steven. "Using Implementation Research to Solve Implementation Problems: The Case of Emergency Energy Assistance." *Journal of Policy Analysis and Management* 4 (Fall 1984): 75–91.

Kentucky Youth Advocates. *Kentucky Complete Streets Toolkit for Advocates.* Prepared by Andrea Plummer (2011). http://kyyouth.org/wp-content/uploads/2013/09/KY-Complete-Streets-Toolkit.pdf.

Kohl, Harold W., III, and Heather Cook, eds *Educating the Student Body: Taking Physical Activity and Physical Education to School.* Washington, D.C.: National Academies Press, October 30, 2013.

Kimmelman, Michael. "Challenging Mayor de Blasio Over Times Square Plazas." *New York Times*, August 21, 2015, C2.

Kingdon, John. *Agendas, Alternatives, and Public Policy.* 2nd ed. London: Longman, 2011.

Klayko, Branden. "Does Move Louisville Have Legs? A Critical Analysis of Louisville's 20-Year Transportation Plan." *Broken Sidewalk*, April 15, 2016.

Klein, Rudolf. *The New Politics of the NHS.* Oxford: Radcliffe, 2010.

Klinenberg, Eric. *Going Solo: The Extraordinary Rise and Surprising Appeal of Living Alone.* London: Penguin, 2012.

Kraft, M. Katherine, and Lawrence D. Brown. "Active Living by Design as a Political Project: Challenges at Three Levels." *American Journal of Preventive Medicine* 37, no. 6 (2009): 453–54.

Kuper, Simon. "Paris in 2050—from Great City to New Metropolis." *Financial Times*, March 14–15, 2020, 19.

Laugesen, Miriam J. *Fixing Medical Prices: How Physicians Are Paid.* Cambridge, Mass.: Harvard University Press, 2016.

Learn-Andes, Jennifer. "Work Begins on River Street Re-Design." *Wilkes-Barre Times Leader*, June 16, 2015, 1A, 4A.

Leighton, Thomas M. "I Believe." Speech, Wilkes-Barre, Pennsylvania, June 9, 2005.

Lewis, Edward. "Gun Violence Shoots Up 21 Percent in Wilkes-Barre." *Wilkes-Barre Times Leader*, July 11, 2013.

Lewis, Michael. *The Fifth Risk.* New York: Norton, 2018.

Lewis, Tom. *Divided Highways: Building the Interstate Highways, Transforming American Life.* New York: Penguin, 1997.

Liebling, A. J. *Between Meals: An Appetite for Paris.* New York: Farrar, Straus and Giroux, 1986.

Lipsky, Michael. *Street-Level Bureaucracy: Dilemmas of the Individual in Public Services.* New York: Russell Sage, 2010.

Litt, Jill S., Hannah L. Reed, Rachel G. Tabak, Susan G. Zieff, Amy A. Eyler, Rodney Lyn, Karin Valentine Goins, Jeanette Gustat, and Nancy O'Hara Tompkins. "Active Living Collaborative in the United States: Understanding Characteristics, Activities, and Achievement of Environmental and Policy Change." *Preventing Chronic Disease* 10, (2013). https://doi.org/10.5888/pcd10.120162.

Louisville-Jefferson County Metro Government. Ordinance No. 128, Series 2019: An Ordinance Amending Ordinance 15, Series 2008, The Complete Streets Policy and Adopting the Policy as an Amendment to Plan 2040, The Comprehensive Plan (Case No. 19AREAPLAN0001). Louisville, Kentucky. https://louisvilleky.gov/document/completestreetsordinance1282019pdf.

Louisville Metro Department of Economic Growth and Innovation. *Vision Louisville Phase 2 Report.* (Louisville, Ky., December 2013). https://louisvilleky.gov/document/vision-phase-2-reportpdf.

Lowi, Theodore J. "American Business, Public Policy, Case Studies, and Political Theory." *World Politics* 16, no. 4 (1964): 677–715.

——. *The End of Liberalism: Ideology, Policy, and the Crisis of Public Authority.* New York: Norton, 1969.

——. "Four Systems of Policy, Politics, and Choice." *Public Administration Review* 32, no. 4 (1972): 298–310.

Lynch, Kevin. *The Image of the City.* Cambridge, Mass.: MIT Press, 1960.

Maarse, Hans, Patrick Jeurissen, and Dirk Ruwaard. "Results of the Market-Oriented Reform in the Netherlands: A Review." *Health Economics, Policy, and Law* 11 (April 2016): 167–78.

MacDonald, John, Charles Branas, and Robert Stokes. *Changing Places: The Science and Art of New Urban Planning.* Princeton, N.J.: Princeton University Press, 2019.

Mann, Ted. "New PAC to Back Pedestrians, Bikes." *Wall Street Journal*, April 25, 2013, A17.

——. "Street Safety Gains Traction as Political Issue." *Wall Street Journal*, August 7, 2013, A17.

Matland, Richard E. "Synthesizing the Implementation Literature: The Ambiguity-Conflict Model of Policy Implementation." *Journal of Public Administration* 5 (1995): 145–74.

Maxey, Heather, Sandra Bishop-Josef, and Ben Goodman. *Unhealthy and Unprepared: National Security Depends on Promoting Healthy Life-Styles from an Early Age*. Washington, D.C.: Council for a Strong America, October 2018.

McCann, Barbara. "Complete the Streets for Smart Growth." *On Common Ground*, June 1, 2007. https://icma.org/documents/complete-streets-smart-growth.

McKeever, Mike. "Sacramento Region Develops New Paradigm for Transportation Planning." Prepared for Presentation to U.S. Senate Committee on Environment and Public Works, Field Briefing, California State Capitol. September 8, 2008.

McLaughlin, Milbrey Wallin. "Learning from Experience: Lessons from Policy Implementation." *Educational Evaluation and Policy Analysis* 9 (1987): 171–78.

McNeill, William H. *Keeping Together in Time: Dance and Drill in Human History*. Cambridge, Mass.: Harvard University Press, 1995.

Merton, Robert. "Bureaucratic Structure and Personality." *Social Forces* 17 (1940): 560–68.

Michener, Jamila. *Fragmented Democracy: Medicaid, Federalism, and Unequal Politics*. New York: Cambridge University Press, 2018.

Mintzberg, Henry. *Managing the Myths of Health Care*. Oakland, Calif.: Bennett-Koehler, 2017.

Moore, Marianne. *The Monkey Puzzler, in Observations*. New York: Farrar, Straus and Giroux, 2016.

Morales, Jeff. "California Blueprint for Bicycling and Walking: A Report to the Legislature." Report to the Legislature, Prepared Pursuant to the Supplemental Report of the 2001 Budget Act, May 2002. https://dot.ca.gov/-/media/dot-media/programs/safety-programs/documents/ped-bike/f0018148-california-blueprint-bicycling-walking-report-a11y.pdf.

Morgan, Kimberly J., and Andrea Louise Campbell. *The Delegated Welfare State: Medicare, Markets and the Governance of Social Policy*. New York: Oxford University Press, 2011.

Moynihan, Daniel Patrick. *Maximum Feasible Misunderstanding: Community Action in the War on Poverty*. New York: Free Press, 1969.

Mueller, Benjamin. "Weekday Car Traffic to Be Banned in Parts of Central and Prospect Parks." *New York Times*, June 19, 2015, A19.

Muller, Jerry Z. *The Tyranny of Metrics*. Princeton, N.J.: Princeton University Press, 2018.

Murphy, Jerome. "Title I of ESEA: The Politics of Implementing Federal Education Reform," *Harvard Educational Review* 41 (April 1971): 35–63.

Nathan, Richard P. *Turning Promises into Performance: The Management Challenges of Implementing Workfare*. New York: Columbia University Press, 1993.

National Center for Health Statistics. *Health, United States, 2012: With Special Feature on Emergency Care*. Hyattsville, MD: National Center for Health Statistics, 2013.

Neenan, William B. *Political Economy of Urban Areas*. Chicago: Markham, 1972.

New York City Department of Education. "Connect with School Wellness Programs." N.d., accessed December 30, 2020. https://infohub.nyced.org/in-our-schools/programs/health-and-wellness/connect-with-school-wellness-programs.

New York City Department of Transportation. *NYC Bike Share Designed by New Yorkers*. N.d., https://www.nyc.gov/html/dot/downloads/pdf/bike-share-outreach-report.pdf.

New York City Obesity Task Force. *Reversing the Epidemic: The New York City Obesity Task Force Plan to Prevent and Control Obesity*, May 31, 2012. http://www.nyc.gov/html/om/pdf/2012/otf_report.pdf.

NYC Parks: Framework for an Equitable Future: Program Overview 2014–2015 (2015). https://healthyplacesbydesign.org/wp-content/uploads/2015/05/CPI-Program-Summary_Updated.pdf.

New York State Senate. "Section 803: Instruction in Physical Education and Kindred Subjects." https://www.nysenate.gov/legislation/laws/EDN/803.

Neustadt, Richard. *Presidential Power and the Modern Presidents: The Politics of Leadership*. New York: Wiley, 1960.

Nir, Sarah Maslin. "New York City's Pedestrian Deaths Hit Lowest Level Since 1910." *New York Times*, January 8, 2018.

O'Hara, John. *Appointment in Samara*. New York: Vintage, 1934.

Oliver, Adam. *The Origins of Behavioral Public Policy*. Cambridge: Cambridge University Press, 2017.

Oliver, Adam, and Lawrence D. Brown. "A Consideration of User Financial Incentives to Address Health Inequalities." *Journal of Health Politics, Policy and Law* 37, no. 2 (2012): 201–26.

Oliver, Eric J. *Fat Politics: The Real Story Behind America's Obesity Epidemic*. New York: Oxford University Press, 2006.

Parilla, Joseph, Sifan Liu, and Marek Gootman. "Charting a Course to the Sacramento Region's Future Economic Prosperity." Brookings Institution, April 30, 2018. https://www.brookings.edu/wp-content/uploads/2018/04/sacramentoregioneconomicprosperity_fullreport.pdf.

Pawel, Miriam. "What Makes California Politics So Special?" *New York Times*, August 19, 2018, 4.

Pear, Robert. "Collecting Disability? Uncle Sam May Not Be a Friend on Social Media." *New York Times*, March 11, 2019, A13.

Peirce, Charles Sanders. *The Collected Works of Charles Sanders Pierce*, ed. Charles Hartshorne and Paul Weiss. Vol. 8, *Reviews, Correspondence, and Bibliography*, Sec. 12. Cambridge, Mass.: Harvard University Press, 1958.

Pessin, Alain. *The Sociology of Howard Becker: Theory with a Wide Horizon*. Chicago: University of Chicago Press, 2017.

Pollack, Andrew. "AMA Recognizes Obesity as a Disease." *New York Times*, June 19, 2013, N1.

Pressman, Jeffrey, and Aaron Wildavsky. *Implementation: How Great Expectations in Washington Are Dashed in Oakland*. Berkeley: University of California Press, 1973.

Proust, Marcel. *A la recherche du temps perdu: Le temps retrouve*. Paris: Gallimard, 1989.

Ramey, Corinne. "Green Space and Health Linked in Vision for South Bronx." *Wall Street Journal*, September 21, 2015, A15–A16.

Robert Wood Johnson Foundation. *Active Living by Design: An RWJF National Program*. Princeton, N.J.: RWJF, last updated February 4, 2013. https://www.rwjf.org/content/dam/farm/reports/program_results_reports/2013/rwjf71184.

Roll, Stephen, Stephanie Moulton, and Jodi Sandfort. "A Comparative Analysis of Two Streams of Implementation Research." *Journal of Public and Nonprofit Affairs* 3 (2017): 3–22.

Rosenblatt, Helena. *The Lost History of Liberalism: From Ancient Rome to the Twenty-First Century*. Princeton, N.J.: Princeton University Press, 2018.

Rykwert, Joseph. *The Seduction of Place: The History and Future of the City*. New York: Vintage, 2002.

Sacramento Area Council of Governments. *2016 Metropolitan Transportation Plan/Sustainable Communities Strategy*. February 18, 2016. www.sacog.org/2016-mtpscs.

Sacramento County Board of Supervisors. *Sacramento County General Plan*. November 9, 2011, https://planning.saccounty.net/PlansandProjectsIn-Progress/Pages/GeneralPlan.aspx.

Sacramento County Transportation. *Bikeway Master Plan*. April 2011. https://sacdot.saccounty.net/Pages/BikewayMasterPlan.aspx.

Sadik-Khan, Janette, and Seth Solomonow. *Street Fight*. New York: Penguin, 2016.

Saetren, Harald. "Facts and Myths About Research on Public Policy Implementation: Out-of-Fashion, Allegedly Dead, But Still Very Much Alive and Relevant." *Policy Studies Journal* 33 (2005): 559–82.

Schasberger, Michele G., Carol S. Hussa, Michael F. Polgar, Julie A. McMonagle, Sharon J. Burke, and Andrew J. Gegaris Jr. "Promoting and Developing a Trail Network Across Suburban, Rural, and Urban Communities." *American Journal of Preventive medicine* 37, no. 6 (December 2009): 336–44.

Schasberger, Michele G., Jessica Raczkowki, Lawrence Newman, and Michael F. Polgar. "Using a Bicycle-Pedestrian Count to Assess Active Living in Downtown Wilkes-Barre." *American Journal of Preventive Medicine* 43 (2012): 399–402.

Schattschneider, E. E. *Politics, Pressures and the Tariff: A Study of Free Private Enterprise in Pressure Politics, As Shown in the 1929–1930 Revision of the Tariff.* New York: Prentice Hall, 1935.

Scott, James C. *Seeing Like a State: How Certain Schemes to Improve the Human Condition Have Failed.* New Haven, Conn.: Yale University Press, 1998.

——. *Two Cheers for Anarchism.* Princeton, N.J.: Princeton University Press, 2012.

Selznick, Philip. *Leadership in Administration.* Evanston, Ill.: Row, Peterson, 1957.

Siciliano, Michael D., Nienke M. Moolenaar, Alan J. Daly, and Yi-Hwa Liou. "A Cognitive Perspective on Policy Implementation: Reform Beliefs, Sensemaking, and Social Networks." *Public Administration Review* 77 (2017): 889–901.

Siedentop, Daryl. *Introduction to Physical Education, Fitness, and Sports*, 7th ed. New York: McGraw-Hill Higher Education, 2009.

Simonet, Daniel. "The New Public Management Theory in the British Health Care System: A Critical Review." *Administration and Society* 47 (2015): 802–26.

Skocpol, Theda. *Social Policy in the United States: Future Possibilities in Historical Perspective* (Princeton, N.J.: Princeton University Press, 1995).

Smith, David G., and Judith D. Moore. *Medicaid Politics and Policy: 1965–2007*. Piscataway, N.J.: Transaction, 2009.

Sparer, Michael S. "Federalism and the Patient Protection and Affordable Care Act of 2010: The Founding Fathers Would Not Be Surprised." *Journal of Health Politics, Policy and Law* 36 (June 2011): 461–68.

——. *Medicaid and the Limits of State Health Reform*. Philadelphia: Temple University Press, 1996.

"Special Issue: Active Living, the Built Environment, and the Policy Agenda." *Journal of Health Politics, Policy and Law* 33 (2008).

Spillane, James P., Brian J. Reiser, and Todd Reimer. "Policy Implementation and Cognition: Reframing and Refocusing Implementation Research." *Review of Educational Research* 72 (2002): 387–431.

Stasi, Linda. "Bloomy the First Boob." *New York Daily News*, August 23, 2015, 17.

Stewart, Matthew. *The Management Myth: Debunking Modern Business Philosophy*. New York: Norton, 2009.

Stockman, David. "The Social Pork Barrel." *The Public Interest* 39 (Spring 1975): 3–30.

Stone, Clarence N. *Regime Politics: Governing Atlanta 1946–1988*. Lawrence: University Press of Kansas, 1989.

Stone, Deborah. *The Disabled State*. Philadelphia: Temple University Press, 1984.

——. *Policy Paradox: The Art of Political Decision Making*, 2nd ed. New York: Norton, 1997.

Stringer, Scott M. "Dropping the Ball: Disparities in Physical Education in New York City's Schools." New York City Comptroller, May 5, 2015. https://comptroller.nyc.gov/reports/dropping-the-ball-disparities-in-physical-education-in-new-york-city/.

Summers, C., L. Cohen, A. Havusha, F. Sliger, and T. Farley. *Take Care New York 2012: A Policy for a Healthier New York City*. New York City Department of Health and Mental Hygiene, September 2009.

Tangle, Andrew. "Bike Lanes Make Cycling in NYC Less Challenging." *Wall Street Journal*, October 9, 2015, H15.

Tonkus, Fran. *Space, the City and Social Theory*. Walden, Mass.: Polity, 2005.

Transportation Alternatives. *The Vision Zero Investment: Why New York Must Rebuild Its Most Dangerous Streets Now*. March 19, 2015. https://static1.squarespace.com/static/5cab9d9b65a707a9b36f4b6c/t/609a8fd546baef1013c28b58/1620742114768/The+Vision+Zero+Investment+-+March+2015.pdf.

United States Department of Health and Human Services. *2008 Physical Activity Guidelines for Americans: Be Active, Healthy, and Happy!* Washington, D.C.: U.S. Department of Health and Human Services, 2008.

United States Department of Health and Human Services. *Physical Activity and Health: A Report of the U.S. Surgeon General, Executive Summary*. Washington, D.C.: U.S. Department of Health and Human Services, 1996.

Vanderbilt, Tom. *Traffic: Why We Drive the Way We Do*. New York: Vintage, 2009.

Vesselinovitch, Andrew. "Pedal Politics." *New York Times*, August 20, 2006.

Vogel, David. *California Greenin': How the Golden State Became an Environmental Leader*. Princeton, N.J.: Princeton University Press, 2018.

Voice of the People. "Readers Sound Off on Mayor Bloomberg and Bicycles and Sen. Ted Cruz." *New York Daily News*, May 5, 2013, 34.

Wagenaar, Hendrik. "'Knowing' the Rules: Administrative Work as Practice." *Public Administration Review* 64 (2004): 643–55.

Walfoort, Nina L., Jennifer J. Clark, Marigny J. Bostock, and Kathleen O'Neil. "Active Louisville: Incorporating Active Living Principles into Planning and Design." *American Journal of Preventive Medicine* 37 (2009): 368–76.

Warner, Sam Bass, and Andrew Whittemore. *American Urban Form: A Representative History*. Cambridge, Mass.: MIT Press, 2012.

Watt, Susan, Wendy Sword, and Paul Krueger. "Implementation of a Health Care Policy: An Analysis of Barriers and Facilitators to Practice Change." *BMC Health Services Research* 5 (2005): 1–10.

Weick, Karl E. *Sensemaking in Organizations*. Thousand Oaks, Calif.: Sage, 1995.

Wennberg, John E. *Tracking Medicine: A Researcher's Quest to Understand Health Care*. New York: Oxford University Press, 2010.

Wilson, James Q. *Bureaucracy: What Government Agencies Do and Why They Do It*. New York: Basic Books, 1989.

——. *Political Organizations*. New York: Basic Books, 1973.

Zusman, Edie E., Sara Jensen Carr, Judy Robinson, Olivia Kasirye, Bonnie Zell, William Jahmal Miller, Teri Duarte, et al. "Moving Toward Implementation: The Potential for Accountable Care Organizations and Private-Public Partnerships to Advance Active Neighborhood Design." *Preventive Medicine* 69, Supplement 1 (2014): 98–101.

INDEX

Milton Keynes UK
Ingram Content Group UK Ltd.
UKHW010622210624
444344UK00003B/46